Preface

This book provides an introduction to quantum mechanics which is meant to be suitable for chemistry undergraduates in their first and second years at UK universities. There are two broad approaches that an author can adopt in writing a book on quantum mechanics. The first is to state all the postulates of quantum mechanics at the outset and then apply them to a variety of phenomena. Although this approach is logical, it seems to leave most students feeling rather bewildered initially because the postulates of quantum mechanics appear to be strange and unrelated to the world of atoms and molecules that they have been taught at school. The second approach, and the one adopted in this book, is to introduce the concepts of quantum mechanics gradually, and illustrate them with simple examples. This inevitably means that some phenomena are discussed before a full explanation of all their aspects can be given. It also means that some examples require a qualitative knowledge of material that is not covered until much later in the book. An example would be the use of π electrons in conjugated hydrocarbons to illustrate the application of the simple particle-in-a-box model. In this book it is assumed that students will already have a qualitative understanding of the general features of atomic and molecular orbitals, probably from attending an introductory course on chemical bonding which is frequently given in first year. The book is therefore intended to be used in conjunction with a lecture course on quantum mechanics given either towards the end of the first undergraduate year, or in the second year, after introductory courses on bonding have been given.

I would like to thank Professor David Phillips for his invitation to write this textbook and for his encouragement during its execution. I would also like to thank Dr Karol Senkiw for producing some of the figures displaying mathematical functions, and for his helpful comments on parts of the text. Finally, I would like to thank my wife, Heather, for her patience during the long hours spent writing this textbook.

David O. Hayward

TUTORIAL CHEMISTRY TEXTS

EDITOR-IN-CHIEF

Professor E W Abel

EXECUTIVE EDITORS

Professor A G Davies
Professor D Phillips
Professor J D Woollins

EDUCATIONAL CONSULTANT

Mr M Berry

This series of books consists of short, single-topic or modular texts, concentrating on the fundamental areas of chemistry taught in undergraduate science courses. Each book provides a concise account of the basic principles underlying a given subject, embodying an independent-learning philosophy and including worked examples. The one topic, one book approach ensures that the series is adaptable to chemistry courses across a variety of institutions.

TITLES IN THE SERIES

Stereochemistry *D G Morris*
Reactions and Characterization of Solids
  S E Dann
Main Group Chemistry *W Henderson*
d- and f-Block Chemistry *C J Jones*
Structure and Bonding *J Barrett*
Functional Group Chemistry *J R Hanson*
Organotransition Metal Chemistry *A F Hill*
Heterocyclic Chemistry *M Sainsbury*
Atomic Structure and Periodicity *J Barrett*
Thermodynamics and Statistical Mechanics
  J M Seddon and J D Gale
Basic Atomic and Molecular Spectroscopy
  J M Hollas
Organic Synthetic Methods *J R Hanson*
Aromatic Chemistry *J D Hepworth,*
  D R Waring and M J Waring
Quantum Mechanics for Chemists
  D O Hayward

FORTHCOMING TITLES

Mechanisms in Organic Reactions
Molecular Interactions
Reaction Kinetics
Lanthanide and Actinide Elements
Maths for Chemists
Bioinorganic Chemistry
Chemistry of Solid Surfaces
Biology for Chemists
Multi-element NMR
EPR Spectroscopy
d-Block Spectroscopy and Magnetism
Biophysical Chemistry
Peptides and Proteins
Natural Products: The Secondary Metabolites

Further information about this series is available at www.rsc.org/tct

Orders and enquiries should be sent to:
Sales and Customer Care, Royal Society of Chemistry, Thomas Graham House,
Science Park, Milton Road, Cambridge CB4 0WF, UK

Tel: +44 1223 432360; Fax: +44 1223 426017; Email: sales@rsc.org

TUTORIAL CHEMISTRY TEXTS

14
Quantum Mechanics for Chemists

DAVID O. HAYWARD

Imperial College of Science, Technology and Medicine, University of London

RS•C

ROYAL SOCIETY OF CHEMISTRY

Cover images © Murray Robertson/visual elements 1998–99, taken from the
109 Visual Elements Periodic Table, available at www.chemsoc.org/viselements

ISBN 0-85404-607-0

A catalogue record for this book is available from the British Library

Published by The Royal Society of Chemistry, Thomas Graham House, Science Park, Milton Road, Cambridge CB4 0WF, UK
Registered Charity No. 207890
For further information see our web site at www.rsc.org

Typeset in Great Britain by Wyvern 21, Bristol
Printed and bound by Polestar Wheatons Ltd, Exeter

Contents

1

Particle–Wave Duality

Aims

By the end of this chapter you should be able to:

- Explain the photoelectric and Compton effects in terms of the particle properties of photons
- Interpret simple photoelectron spectra and calculate the binding energies of the electronic states from which the photoelectrons arise
- Use the de Broglie relationship to calculate the wavelengths of beams of electrons, atoms and molecules
- Interpret diffraction phenomena, in particular low-energy electron diffraction
- Write down the wavefunction for a particle moving with constant momentum

1.1   Introduction

In everyday life a clear distinction is made between particles and waves. A particle has a well-defined mass, and its position and velocity can be accurately determined as a function of time by applying Newton's laws of motion. Waves do not have mass in the normal sense and cannot be precisely located; they are best described in terms of a characteristic frequency and wavelength. Waves also have the important property that they can interact with one another to produce interference patterns, a process known as diffraction. An example of this is shown in Figure 1.1, where interference is observed between two synchronous wave sources. When we look at the interaction of electrons with a metal surface we find that they can be diffracted in a similar way, and it must be concluded that they, too, have wave-like properties, although they are normally thought of as particles. Similarly, electromagnetic radiation can show particle-like properties, an example being the way in which X-rays can knock electrons out of solids (see Figure 1.2). This process can be understood in terms of the collision of two particles, the X-ray particles (photons) having a momentum inversely proportional to their wavelength.

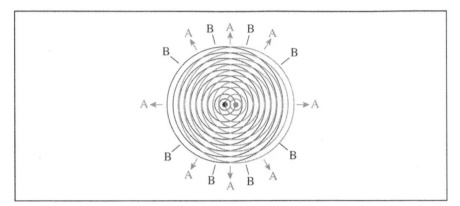

Figure 1.1 Circular waves emanating from two synchronized sources. The waves reinforce one another in the directions marked A but cancel one another out in the directions marked B

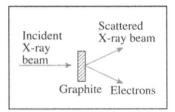

Figure 1.2 Experiment that demonstrates the particle properties of X-rays

Although recognition of the particle–wave duality of basic entities such as light and electrons is a product of 20th century science, there had been earlier signs of a problem. During the preceding centuries there had been a disagreement about the true nature of light, with the followers of Newton favouring a corpuscular theory, whereas the followers of Huygens regarded light as a wave motion. This difference of opinion appeared to have been settled by James Young in 1801, when he showed that interference patterns could be produced when light was passed through two closely spaced slits. This result could be explained satisfactorily only if light was a wave motion. Nowadays, it is accepted that both points of view have some validity, light possessing some of the properties of both a wave and a particle. The same is true of electrons, atoms and molecules. However, it would be wrong to infer from this that there is no fundamental difference between particles and electromagnetic waves; one major difference is that electromagnetic waves always move (with respect to the observer) at the velocity of light (3×10^8 m s^{-1} in vacuum), whereas particles can be observed at rest.

The wave properties of matter can also become evident in ways other than by diffraction. An atom in a solid will vibrate because of its thermal motion. In classical mechanics the energy resulting from this vibrational motion can have a continuous range of values, but this is no longer true when the wave properties of the atom are taken into account. An imperfect analogy would be with the modes of vibration of a string, which are illustrated in Figure 1.3. For a string of length L, vibrations can be generated with wavelengths equal to $2L$, $2L/2$, $2L/3$, *etc.* These correspond to the fundamental vibration and its overtones. A similar process occurs with the vibrating atom, which can be considered to vibrate within the confines of a potential well. Only matter waves with particular wavelengths will fit into this well, and they correspond to discrete energies which, for simple harmonic motion, are given by the formula:

$$E = \left(n + \frac{1}{2}\right) h \omega_{vib} \qquad (1.1)$$

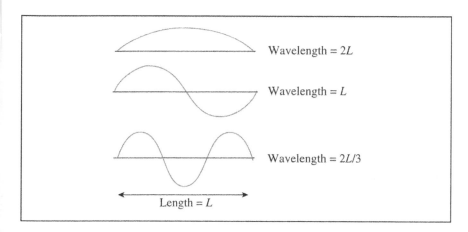

Wavelength = 2L

Wavelength = L

Wavelength = 2L/3

Length = L

Figure 1.3 Some of the standing waves that can be generated on a string

Here, n is an integer, ω_{vib} is the frequency of the vibration, and h is known as Planck's constant. This quantization of energy, as it is known, was first postulated by Max Planck in 1900 as a key part of his theory to explain the frequency distribution of radiation emitted by a black body. It is found that energy is quantized whenever a particle is confined to a small space because of the need to match the wavefunction of the particle to the space available. This applies just as much to electrons travelling around an atomic nucleus as it does to atoms vibrating in a solid.

The major part of this book will be concerned with the wave properties of matter, but it will be helpful, at the outset, to spend a little time looking at the particle properties of electromagnetic radiation because similar concepts apply in both cases.

1.2 Particle Properties of Electromagnetic Waves

1.2.1 The Photoelectric Effect

When a metal is placed in an evacuated chamber and illuminated with ultraviolet (UV) light, electrons are emitted (see Figure 1.4). These are known as photoelectrons, and they flow from the metal surface exposed to the radiation to the collecting electrode, where the current flowing can be measured with a sensitive galvanometer. This phenomenon was first studied systematically by P. Lenard who, in 1900, was able to show that the charged particles emitted from the metal surface were electrons.

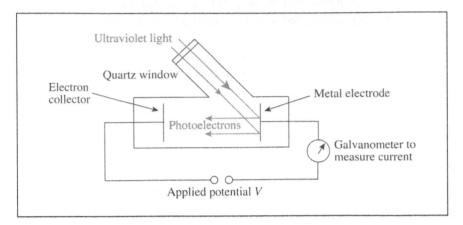

Figure 1.4 Schematic drawing of apparatus used by Lenard to observe the photoelectric effect

By varying the applied potential V it was possible to find the value V_0 (a retarding potential, with the collecting electrode negative with respect to the metal under study) at which the current went to zero. The value of V_0 was found to vary with the frequency of the radiation used, as shown in Figure 1.5. When the potential difference was equal to V_0, even the most energetic electrons were not quite able to reach the collector, and the maximum kinetic energy of the photoelectrons must therefore have been equal to eV_0, where e stands for the charge on one electron.

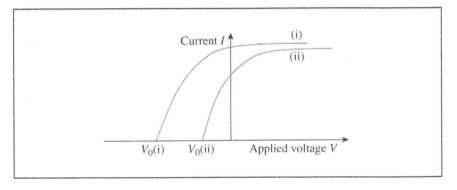

Figure 1.5 Variation of the photoelectric current I with the potential difference V applied between the electrodes when illuminated with light of two different frequencies. The frequency of the light would be higher for curve (i) than for curve (ii)

In a later series of experiments, conducted between 1914 and 1916, R. A. Millikan was able to show that the maximum kinetic energy of the photoelectrons varied linearly with the frequency of the light used, and that the slope of the plots of maximum kinetic energy versus frequency was equal to Planck's constant h. A schematic representation of Millikan's results is shown in Figure 1.6.

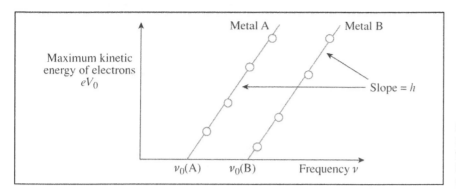

The results can be summarized as follows:

1. For any particular metal there was a critical radiation frequency v_0 below which no electrons were emitted, no matter what value of the retarding potential V was used.
2. The maximum kinetic energy of the emitted electrons, $\frac{1}{2}m(v_{max})^2$ or eV_0, depended on the frequency but not the intensity of the light used.
3. The slope of the plot of maximum kinetic energy versus frequency was the same for all metals, with a value equal to Planck's constant h.

In 1905, Einstein had already postulated that radiant energy came in small packets, each with an energy equal to hv, where v is the frequency of the radiation, and Millikan's results gave strong support to this concept. Part of the energy is used to release the electron from the metal and the rest appears as the kinetic energy of the electron. Electrons can be released from different energy levels, the highest occupied level in a metal being known as the Fermi level (see Figure 1.7). The minimum energy required to release an electron from the Fermi level is known as the work function of the metal and is given the symbol ϕ. Clearly those electrons released from the Fermi level will have the highest kinetic energy. This can be expressed in mathematical form as:

$$hv = T + \phi \qquad (1.2)$$

where T is the maximum kinetic energy of the emitted electrons. This phenomenon shows that light can behave as a particle (that is, a photon) as well as a wave.

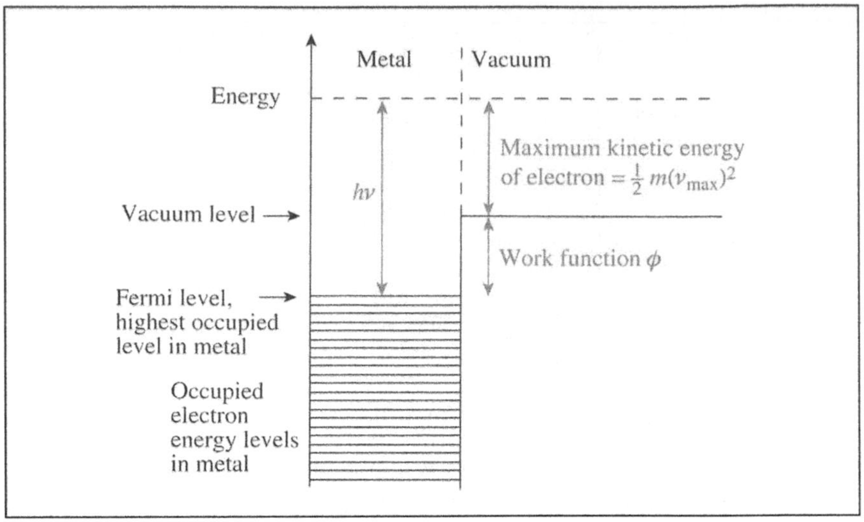

Figure 1.7 Electron energy level diagram for a metal, showing the relationship between the photon energy, the work function of the metal, and the maximum amount of kinetic energy that a photoelectron can possess

One **electronvolt** is the energy that an electron gains when it is accelerated through a potential difference of one volt.

Worked Problem 1.1

Q The longest wavelength of light that will induce emission of electrons from potassium is 564 nm. Calculate the work function of potassium in units of Joules and also in units of electronvolts.

A At the critical frequency, v_0, photoelectrons are emitted with zero kinetic energy and equation (1.2) then becomes $hv_0 = \phi$. The wavelength of the light, λ_0, multiplied by the frequency, v_0, is equal to the velocity of light, c. Thus:

$$\phi = \frac{hc}{\lambda_0} = \frac{(6.626 \times 10^{-34} \text{ J s}) \times (2.998 \times 10^8 \text{ m s}^{-1})}{564 \times 10^{-9} \text{ m}} = 3.52 \times 10^{-19} \text{ J}$$

The work function is more conveniently given in units of electronvolts. Since the electronic charge is equal to 1.602×10^{-19} Coulomb:

$$1 \text{ eV} = (1.602 \times 10^{-19} \text{ C}) \times (1 \text{ V}) = 1.602 \times 10^{-19} \text{ J}$$

Thus:

$$\phi = \frac{3.52 \times 10^{-19} \text{ J}}{1.602 \times 10^{-19} \text{ J (eV)}^{-1}} = 2.20 \text{ eV}$$

1.2.2 Photoelectron Spectroscopy

The photoelectric effect forms the basis of the modern technique of photoelectron spectroscopy[1] in which molecules are irradiated with UV light and the kinetic energies of the emitted photoelectrons are measured with an electrostatic energy analyser. A schematic diagram of the apparatus used is shown in Figure 1.8. This technique provides valuable information about the binding energies of electrons in molecules. Unlike metals, molecules have discrete energy levels, and equation (1.2) can be rewritten as:

$$h\nu = T + I \tag{1.3}$$

where I is the ionization energy of the electron that has been emitted and is used in place of the work function ϕ.

The **ionization energy** (strictly, the first ionization energy) is the energy required to remove an electron from a neutral gaseous molecule (M) according to the process:
$$M(g) \rightarrow M^+(g) + e^-$$

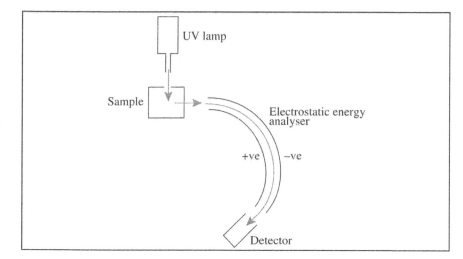

Figure 1.8 Schematic diagram of the apparatus used to study the UV photoelectron spectra of gases

Thus, by irradiating the sample with light of known frequency ν and measuring the kinetic energy of the photoelectrons emitted, the ionization energy I can be determined. The UV light is often provided by a helium discharge lamp, which generates photons with a characteristic energy.

The photoelectrons may originate from a variety of electronic states within the molecule, and each state will have a different ionization energy, I. Thus, when the photoelectric current is plotted as a function of the kinetic energy of the photoelectrons, a series of peaks will be obtained. A schematic diagram of the photoelectron spectrum of nitrogen, excited by helium-I radiation, is shown in Figure 1.9. Here, the data have been plotted both as kinetic energies of the photoelectrons and as ionization energies. The process occurring can be represented as:

$$h\nu + N_2 \rightarrow N_2^+ + e^- \tag{1.4}$$

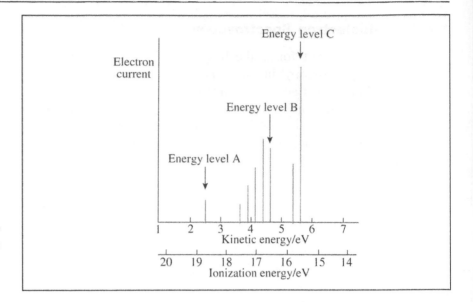

Figure 1.9 Schematic diagram of the photoelectron spectrum of nitrogen. The arrows indicate the energies of photoelectrons that are emitted without vibrational excitation

Photoelectrons can be emitted from three different energy levels in the nitrogen molecule, labelled A, B and C. The other peaks appearing in the spectrum are caused by vibrational excitation of the nitrogen molecule ion that is formed. This reduces the kinetic energy of the photoelectrons because it takes additional energy to make the ion vibrate, and therefore these peaks all appear to the left of the principal peaks.

Worked Problem 1.2

Q Photoelectrons emitted from nitrogen gas exposed to helium-I radiation (wavelength = 58.43 nm) have kinetic energies of 5.63, 4.53 and 2.50 eV. These energies correspond to formation of N_2^+ without vibrational excitation. Calculate the photon energy of the He-I radiation in units of electronvolts, and hence determine the ionization energies of the electron states from which the photoelectrons originate, also in electronvolts.

A
$$\text{Photon energy} = \frac{hc}{\lambda} = \frac{\left(6.626 \times 10^{-34} \text{ J s}\right) \times \left(2.998 \times 10^8 \text{ m s}^{-1}\right)}{\left(5.843 \times 10^{-8} \text{ m}\right)}$$
$$= 3.400 \times 10^{-18} \text{ J}$$

In electronvolts this becomes:

$$E_{\text{photon}} = \frac{3.400 \times 10^{-18}\ \text{J}}{1.602 \times 10^{-19}\ \text{J (eV)}^{-1}} = 21.22\ \text{eV}$$

The ionization energies are therefore:

$$21.22 - 5.63 = 15.59\ \text{eV}$$
$$21.22 - 4.53 = 16.69\ \text{eV}$$
$$21.22 - 2.50 = 18.72\ \text{eV}$$

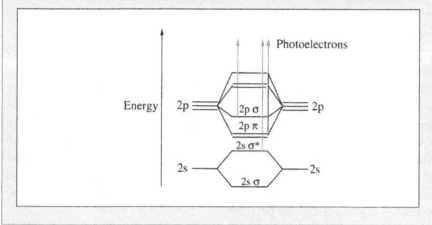

This note is meant for students who already have a qualitative understanding of molecular orbitals, a subject often covered in first year courses in inorganic chemistry.
The largest ionization energy (18.7 eV) is that required to remove an electron from the antibonding 2s molecular orbital of nitrogen. The other two ionization energies (16.7 and 15.6 eV) represent electrons that have originated from the 2p π and 2p σ molecular orbitals. These are shown in Figure 1.10.

Figure 1.10 Diagram showing the molecular orbitals of nitrogen from which the photoelectrons are emitted

1.2.3 Collisions between Photons and Electrons: the Compton Effect

We have seen that light can be treated as particles (photons) with an energy equal to $h\nu$. The energy of the photon can also be obtained from Albert Einstein's famous equation relating mass and energy:

$$E = mc^2 \qquad (1.5)$$

where c is the velocity of light. The mass referred to in this equation is the inertial mass of the photon, which arises as a direct consequence of the photon's energy. It should be noted that the photon does not have a proper, or rest, mass in the way that an electron does. When the two expressions for the energy are equated, we arrive at the equation $mc^2 = h\nu = hc/\lambda$. The momentum, p, of the photon is then given by the equation:

$$p = mc = h/\lambda \qquad (1.6)$$

This formula was verified by Compton in 1923. He found that when a monochromatic beam of X-rays was incident on a block of graphite (see Figure 1.11), two kinds of X-rays emerged on the far side. One had the wavelength of the original beam, λ_i, but the other had a longer wavelength, λ_s.

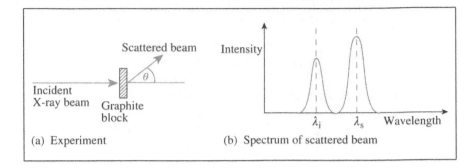

Figure 1.11 The Compton experiment

These results were explained by postulating that some of the X-ray photons had undergone elastic collisions with electrons in the graphite. This process, shown in Figure 1.12, results in the scattered photon having a smaller momentum than the incident photon. The binding energy of the electron in the graphite is very small compared to the energy of an X-ray photon, and the electron behaves as though it was a free electron which is virtually at rest before the collision. By applying the laws of conservation of energy and linear momentum to the collision and assuming that the momentum of an X-ray photon is equal to h/λ, the following equation can be derived:

$$\Delta\lambda = \frac{h}{m_e c}\left(1 - \cos\theta\right) \tag{1.7}$$

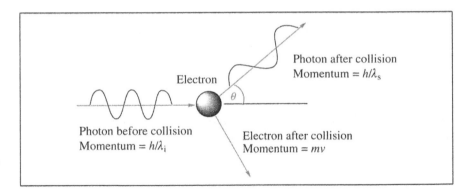

Figure 1.12 Depiction of the collision of an X-ray photon with an electron

Here, $\Delta\lambda$ is the change in wavelength of the photon, m_e is the mass of the electron, and θ is the scattering angle. This equation is in good agreement with the experimental results, thus demonstrating that the photon does indeed have a momentum equal to h/λ.

Worked Problem 1.3

Q Calculate the momentum of an X-ray photon with a wavelength of 0.17 nm. How does this compare with the momentum of an electron that has been accelerated through a potential difference of 100 volts?

A The momentum of the photon is given by the equation:

$$p = \frac{h}{\lambda} = \frac{6.626 \times 10^{-34} \text{ J s}}{0.17 \times 10^{-9} \text{ m}} = 3.9 \times 10^{-24} \text{ kg m s}^{-1}$$

Note on units: by using Newton's second law of motion and the definition of work, it can be shown that $J \equiv \text{kg m}^2 \text{ s}^{-2}$.

In this problem the velocity of the electron is much less than the velocity of light and relativistic corrections to the mass need not be made. Under these circumstances the momentum of the electron, p, can be obtained from the classical equation relating energy and momentum:

$$E = \tfrac{1}{2}mv^2 = \frac{p^2}{2m}$$

where v is the velocity and m is the mass of the electron. Thus:

$$p = \sqrt{2mE}$$

The energy is equal to the electronic charge multiplied by the voltage. Thus:

$$E = (1.602 \times 10^{-19} \text{ C}) \times (100 \text{ V}) = 1.602 \times 10^{-17} \text{ J}$$

and

$$p = \sqrt{2mE} = \sqrt{2 \times (9.109 \times 10^{-31} \text{ kg}) \times (1.602 \times 10^{-17} \text{ J})}$$
$$= 5.40 \times 10^{-24} \text{ kg m s}^{-1}$$

It can be seen that the momenta of the electron and the X-ray photon have similar magnitudes.

1.3   Wave Properties of Matter

1.3.1 De Broglie Waves and Low-energy Electron Diffraction (LEED)

In 1924, Louis de Broglie proposed that all matter has wave properties and that the wavelength of the associated wave is related to the momentum, p, of the particle by the expression already derived for photons, namely:

$$\lambda = h/p \qquad (1.8)$$

In 1927, Davisson and Germer demonstrated that electrons can be diffracted from the surface of a nickel crystal, a result which can be explained only if electrons have wave-like properties. By applying the de Broglie relationship they were able to calculate the interatomic spacing between the nickel atoms, and their result agreed with the value obtained from X-ray diffraction measurements.

Davisson and Germer used an electron beam that had been accelerated through a potential difference of 54 volts and the electrons approached the surface at normal incidence, *i.e.* perpendicularly, as shown in Figure 1.13. The nickel sample had been heated to high temperatures before the experiments began, and this had caused large crystals to form which produced the regular array of nickel atoms shown in the figure. Unlike X-rays, electrons do not penetrate very far into the metal crystal, and most of the scattering occurs from the top layer of metal atoms. Scattering of electrons was found to occur most strongly at an angle of 50° to the surface normal, and this result can be explained in terms of the constructive interference of electron waves scattered from neighbouring nickel atoms, as shown in Figure 1.14.

The condition for constructive interference is that the waves should all be in phase with one another. Figure 1.15 gives examples of waves

Figure 1.13 The arrangement used by Davisson and Germer to observe diffraction of electrons from the surface of a nickel crystal

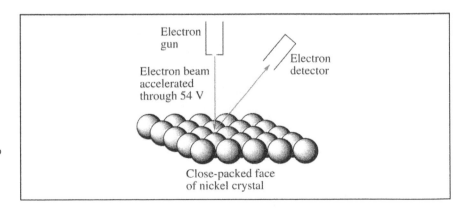

Electron gun

Electron beam accelerated through 54 V

Electron detector

Close-packed face of nickel crystal

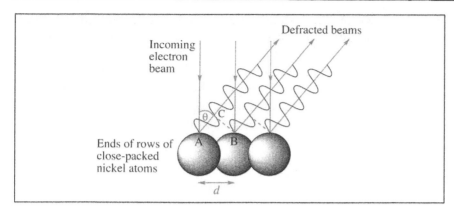

Figure 1.14 Constructive interference of an electron beam at a nickel surface

that interfere constructively and destructively with one another. From an inspection of Figure 1.14, constructive interference occurs when the difference in path lengths, AC, is equal to a whole number of wavelengths. Now AC = $d\sin\theta$, where d is the perpendicular distance between the close-packed rows of nickel atoms. The condition for constructive interference then becomes:

$$d\sin\theta = n\lambda \tag{1.9}$$

where n is an integer. λ can be obtained from the equation $\lambda = h/p$ and hence d can be calculated.

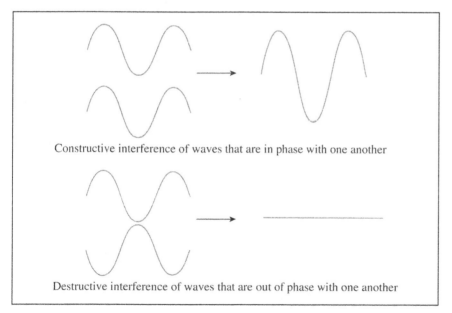

Constructive interference of waves that are in phase with one another

Destructive interference of waves that are out of phase with one another

Figure 1.15 Constructive and destructive interference of waves

Constructive interference occurs only in certain directions. One of these directions is shown in Figure 1.16 for first-order diffraction, where

$n = 1$. Rows of close-packed nickel atoms occur in three different directions and diffraction at right angles to these rows results in the diffraction pattern shown in Figure 1.17, with six first-order diffraction spots. In Davisson and Germer's original experiments, only one diffraction peak was observed but all six were seen in later work. By using the equipment shown diagrammatically in Figure 1.18(a) the diffraction pattern shown in Figure 1.18(b) can be obtained.

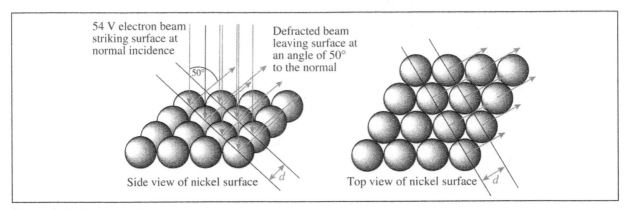

54 V electron beam striking surface at normal incidence

50°

Defracted beam leaving surface at an angle of 50° to the normal

Side view of nickel surface d

Top view of nickel surface d

Figure 1.16 Constructive interference of electrons scattered from a nickel surface in one particular direction

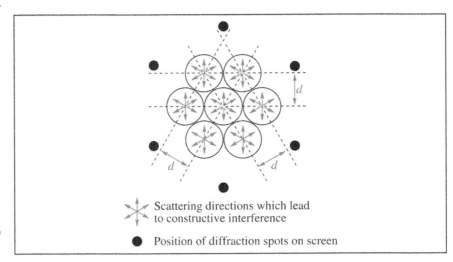

d

d d

Scattering directions which lead to constructive interference

Position of diffraction spots on screen

Figure 1.17 The origins of the first-order diffraction pattern

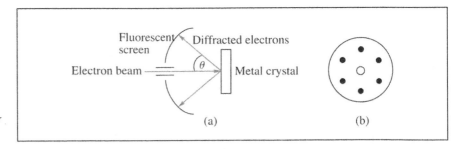

Fluorescent screen   Diffracted electrons

Electron beam θ Metal crystal

(a)   (b)

Figure 1.18 (a) LEED apparatus and (b) the first-order diffraction pattern

Box 1.1 Low-energy Electron Diffraction

Low-energy electron diffraction (LEED) is a very important technique for the study of surface structures formed by atoms and molecules adsorbed on metal and semiconductor surfaces, and it is widely used today in surface science and heterogeneous catalysis.[2]

The sensitivity of the diffraction pattern to the adsorption of gases is illustrated for the adsorption of oxygen. Figure 1.19 shows the ordered structure that is formed when a close-packed nickel surface is exposed to oxygen, and the diffraction pattern that is obtained. The distance between the oxygen atoms is twice the distance between the nickel atoms and the value of d to be used in equation (1.9) is therefore doubled. The spacing between the diffraction spots seen on the screen is proportional to $\sin\theta$, and since $d\sin\theta$ is constant at a given wavelength, the spacing in the diffraction pattern is halved.

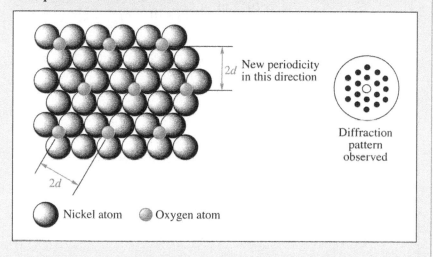

Figure 1.19 The structure of oxygen adsorbed on a close-packed nickel surface together with the corresponding LEED pattern

Worked Problem 1.4

Q In the experiments of Davisson and Germer, an electron beam with an energy of 54.0 eV struck a close-packed nickel surface perpendicularly. A diffracted beam was observed at an angle of 50° to the perpendicular. Calculate (i) the wavelength of the electrons, (ii) the spacing between the rows of nickel atoms and (iii) the metallic radius of nickel.

A (i) The energy of the electron = $(54.0 \text{ V}) \times (1.602 \times 10^{-19} \text{ C}) =$

8.651 × 10^{-18} J. As in Worked Problem 1.3, the momentum, p, of the electron can be found by using the equation:

$$p = \sqrt{2mE} = \sqrt{2 \times (9.109 \times 10^{-31} \text{ kg}) \times (8.651 \times 10^{-18} \text{ J})}$$
$$= 3.97 \times 10^{-24} \text{ kg m s}^{-1}$$

The wavelength of the electron is then calculated from the de Broglie equation:

$$\lambda = \frac{h}{p} = \frac{6.626 \times 10^{-34} \text{ J s}}{(3.97 \times 10^{-24} \text{ kg m s}^{-1})} = 1.67 \times 10^{-10} \text{ m}$$

(ii) The spacing d can be found from the diffraction formula, $n\lambda = d\sin\theta$. Thus:

$$d = \frac{1.67 \times 10^{-10} \text{ m}}{\sin 50°} = 2.18 \times 10^{-10} \text{ m}$$

(iii) The relationship between d and the metallic radius a is illustrated in Figure 1.20. By simple geometry:

$$a = \frac{d}{\sqrt{3}} = 1.26 \times 10^{-10} \text{ m}$$

Figure 1.20 Geometry of a close-packed nickel surface

1.3.2 Diffraction of Atoms and Molecules from Surfaces

Diffraction of helium from alkali halide crystals was first observed by Estermann, Frisch and Stern in 1931, but it was not until the 1970s that diffraction of atoms and molecules from metal surfaces was first clearly demonstrated. A beam of atoms or molecules, all with approximately the same energy, can be generated by expanding gas at high pressure through a supersonic nozzle. With the set-up shown in Figure 1.21 it has been possible to observe the diffraction of beams of helium and deuterium from a close-packed nickel surface.[3] Six first-order diffraction peaks have been observed, and the de Broglie relationship has been used to calculate the spacing between the rows of nickel atoms, just as in the experiments of Davisson and Germer.

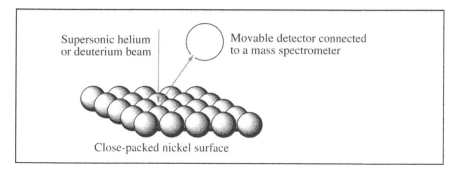

Figure 1.21 Set-up used to observe diffraction of helium or deuterium from a nickel surface

Neutrons can also be diffracted by crystals, but they travel a long way into the solid, and therefore give information on the internal structure of the solid rather than on the structure of the surface.

Worked Problem 1.5

Q A beam of helium atoms, originating from a supersonic nozzle source held at a temperature of 300 K, strikes a close-packed nickel surface perpendicularly. The kinetic energy of the helium atoms is equal to $2.5\, k_B T_{nozzle}$, where k_B is the Boltzmann constant. Given that the spacing between the rows of nickel atoms is 216 pm, calculate the angle to the perpendicular at which the six first-order diffraction peaks will be observed.

A The energy of the helium atom is:
$E = 2.5 \times (1.381 \times 10^{-23}\ \mathrm{J\ K^{-1}}) \times (300\ \mathrm{K}) = 1.036 \times 10^{-20}\ \mathrm{J}$
The mass of a helium atom is obtained from the equation:

$$m = \frac{4.003 \times 10^{-3}\,\mathrm{kg\ mol^{-1}}}{6.022 \times 10^{23}\,\mathrm{mol^{-1}}} = 6.647 \times 10^{27}\,\mathrm{kg}$$

The relationship between momentum, p, and energy is $p = \sqrt{2mE}$. The wavelength can then be found from the de Broglie relationship:

$$\lambda = \frac{h}{p} = \frac{h}{\sqrt{2mE}} = \frac{6.626 \times 10^{-34}\,\mathrm{J\ s^{-1}}}{\left(2 \times (6.647 \times 10^{-27}\,\mathrm{kg}) \times (1.036 \times 10^{-20}\,\mathrm{J})\right)^{1/2}}$$

$$= 5.65 \times 10^{-11}\,\mathrm{m}$$

The diffraction formula is $n\lambda = d\sin\theta$ and $n = 1$ for first-order diffraction. Thus:

$$\sin\theta = \frac{5.65 \times 10^{-11}\,\text{m}}{2.16 \times 10^{-10}\,\text{m}} = 0.262 \text{ and } \theta = 15.2°.$$

1.4   Matter Waves

1.4.1   The One-dimensional Form of the Wavefunction ψ

Mathematical aside: in this equation, i represents the square root of minus one. Since

$$e^x = 1 + x + \frac{x^2}{2!} + \frac{x^3}{3!} + \frac{x^4}{4!} + \frac{x^5}{5!} + \frac{x^6}{6!}\ldots$$

it follows that

$$e^{ix} = 1 + ix - \frac{x^2}{2!} - \frac{ix^3}{3!} + \frac{x^4}{4!} + \frac{ix^5}{5!} - \frac{x^6}{6!}\ldots$$

$$= (1 - \frac{x^2}{2!} + \frac{x^4}{4!} - \frac{x^6}{6!}\ldots) + i(x - \frac{x^3}{3!} + \frac{x^5}{5!}\ldots)$$

The first term in parentheses is the expansion of $\cos x$ and the second term in parentheses is the expansion of $\sin x$. Thus $e^{ix} = \cos x + i\sin x$.

In this section we consider a beam of particles moving along the x axis at constant flux and velocity. It is assumed that the actual positions of particles within the beam at any particular time are unknown. Under these circumstances the de Broglie wavefunction of a particle in the beam will be a function only of x, and the time will not be involved. The simplest wavefunction would be $\psi = A\sin kx$ or $\psi = A\cos kx$, where A is the amplitude of the wave and k is a constant, but, for reasons that will gradually become clear, it has been found that the most effective way of representing a particle moving along the x axis with constant momentum is by the complex wavefunction:

$$\psi = Ae^{ikx} = A(\cos kx + i\sin kx) \tag{1.10}$$

This wavefunction consists of two sinusoidal waves, one real and one complex, which are 90° out of phase with one another (see Figure 1.22).

In three dimensions the constant k becomes a vector, which is often referred to as the wave vector. Here, k will be used to indicate the magnitude of the wave vector, and this can be related to the wavelength λ by noting that the magnitude of the wavefunction ψ does not change

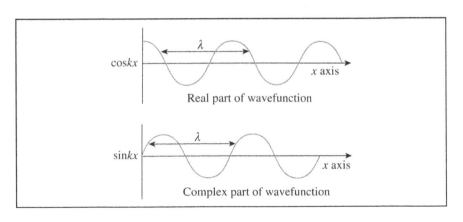

Figure 1.22 Real and complex parts of the wavefunction

when x is increased by an amount equal to λ. Thus, for the sine component:

$$A\sin kx = A\sin k(x + \lambda) = A\sin(kx + k\lambda) \qquad (1.11)$$

and similarly for the cosine component. Now, $\sin\theta = \sin(\theta + 2\pi)$ and $\cos\theta = \cos(\theta + 2\pi)$, from which it follows that $k\lambda$ must be equal to 2π. Thus:

$$k = 2\pi/\lambda \qquad (1.12)$$

According to the de Broglie relationship, $p = h/\lambda$, and therefore:

$$p = \frac{hk}{2\pi} = \hbar k \qquad (1.13)$$

where $\hbar$ is equal to $h/(2\pi)$.

The kinetic energy, T, is equal to $p^2/2m$, and when this is combined with equation (1.13) we obtain:

$$T = (\hbar k)^2/(2m) \qquad (1.14)$$

Thus, the more kinetic energy the particle has, the smaller the wavelength becomes and we can regard the curvature of the wavefunction as giving a measure of the kinetic energy of the particle.

If k is taken to be a positive quantity, then a particle moving in the positive x direction with momentum $\hbar k$ would have a wavefunction $\psi = Ae^{ikx}$, and a particle moving in the opposite (negative x) direction would have a wavefunction $\psi = Ae^{-ikx}$. When expanded in terms of sine and cosine functions, the two wavefunctions become:

$$\psi = Ae^{ikx} = \cos kx + i\sin kx \qquad (1.15)$$

and:

$$\psi = e^{-ikx} = \cos(-kx) + i\sin(-kx) = \cos kx - i\sin kx \qquad (1.16)$$

These wavefunctions are shown in Figure 1.24, where it can be seen that the phase relationship between the real and complex parts is different for motion in opposite directions.

One reason for using the complex wavefunction now becomes clear: it allows motion in opposite directions to be distinguished because the two wavefunctions, $\psi = Ae^{ikx}$ and $\psi = Ae^{-ikx}$, are distinct. If a simple sine or cosine function had been chosen to represent the motion of the electron, it would have been impossible to distinguish between motion

The use of $\hbar$ as a shorthand way of writing $h/(2\pi)$ will be extensively used throughout this book.

Mathematical aside: it can be seen from Figure 1.23 that $\cos(-\theta) = \cos\theta$, and that $\sin(-\theta) = -\sin\theta$.

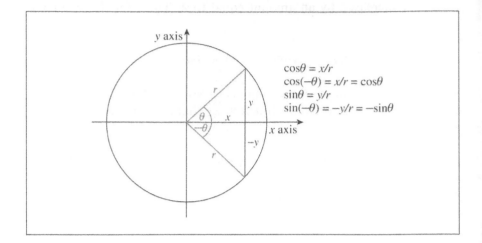

Figure 1.23 The sines and cosines of negative angles

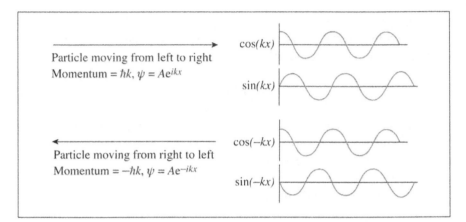

Figure 1.24 Wavefunctions for movement in opposite directions

in opposite directions because $\cos(-kx) = \cos(kx)$ and $\sin(-kx) = -\sin(kx)$. Although the sine function changes sign when k goes to $-k$, this does not produce a new wavefunction because the sign of the wavefunction is arbitrary and no particular physical significance can be attached to it.

Worked Problem 1.6

Q An electron beam is accelerated through a potential difference of V volts and then has a wavelength of 0.3 nm. Calculate values for (a) the potential difference V, (b) the magnitude of the wave vector k and (c) the momentum of an electron in the beam.

A (a) The energy of the electron, E, is equal to eV, where e is the electronic charge. The momentum is then equal to $\sqrt{2meV}$. When this is combined with the de Broglie relationship we obtain $h/\lambda=(2meV)^{1/2}$. After rearrangement this becomes:

$$V = \frac{h^2}{2me\lambda^2} = \left[\frac{6.626\times10^{-34}\text{ J s}}{0.3\times10^{-9}\text{ m}}\right]^2 \times \frac{1}{2\times(9.109\times10^{-31}\text{ kg})\times(1.602\times10^{-19}\text{ C})} = 16.7\text{ V}$$

(b) The magnitude of the wave vector is equal to $2\pi/\lambda$. Thus:

$$|k| = \frac{2\pi}{0.3\times10^{-9}\text{ m}} = 2.09\times10^{10}\text{ m}^{-1}$$

(c) $$p = \frac{h}{\lambda} = \frac{6.626\times10^{-34}\text{ J s}^{-1}}{0.3\times10^{-9}\text{ m}} = 2.21\times10^{-24}\text{ kg m s}^{-1}$$

1.4.2 The Interpretation of the Wavefunction in Terms of Probabilities

It is difficult to give the wavefunction ψ a physical significance because, as we have just seen, it can have both real and complex parts. In 1926, Max Born suggested that the wavefunction of a particle, ψ, multiplied by its complex conjugate, ψ^*, might be connected with the probability of finding the particle at a particular point. We have to be careful here, because the probability of finding the particle at the point x,y,z depends on how accurately the point is defined. The more exact the definition, the lower is the probability of finding the particle there. It is rather like asking how many people in Great Britain are exactly six feet tall. A moment's reflection will show that the question is meaningless unless a height range is specified. A sensible question would be "How many people in Great Britain have a height between six feet and six feet and one inch?" In the same way, we need to find the probability of finding the particle in an infinitesimally small box with dimensions dx,dy,dz (known as a volume element), located at the point x,y,z. This is illustrated in Figure 1.25. The probability of finding the particle within this infinitesimal box is clearly going to be proportional to its volume, and the probability will therefore be given by the equation:

$$P(x,y,z)dxdydz = \psi^*\psi dxdydz \qquad (1.17)$$

where $P(x,y,z)dxdydz$ is the probability of finding the particle between x and $x + dx$, y and $y + dy$ and z and $z + dz$. $P(x,y,z)$ is known as the

Mathematical aside: the complex conjugate of a complex number is obtained by putting a minus sign in front of i, wherever it occurs. Thus, if $\psi = a + ib$, then $\psi^* = a - ib$. When a complex number is multiplied by its complex conjugate, the result is always a real number: $\psi^*\psi = (a + ib)(a - ib) = a^2 + b^2$.

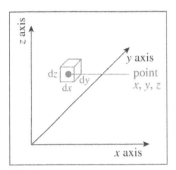

Figure 1.25 The volume element to which the probability refers

probability density. It is equal to $\psi^*\psi$ and has dimensions in SI units of m^{-3}. It follows that the wavefunction itself has a dimension of $m^{-3/2}$.

For motion restricted to the x axis, equation (1.17) becomes:

$$P(x)dx = \psi^*\psi dx \qquad (1.18)$$

and the wavefunction will have a dimension of $m^{-1/2}$.

1.4.3 Application to a Particle Moving with Constant Momentum

For a particle moving along the x axis with constant momentum $\hbar k$, $\psi = Ae^{ikx}$ and $\psi^* = Ae^{-ikx}$. The probability of finding the particle between x and $x + dx$ is then given by the expression:

$$P(x)dx = \psi^*\psi dx = A^2 dx \qquad (1.19)$$

It follows that the particle is equally likely to be found anywhere along the x axis, which is equivalent to stating that its position at any instant is unknown. From this we conclude that a particle with wavefunction $\psi = Ae^{ikx}$ has a definite momentum but an undefined position. We will come back to this subject in Chapter 3, when the Heisenberg Uncertainty Principle is discussed.

1.4.4 Time-independent Wavefunctions

All the wavefunctions that will be considered in this book are functions only of the spatial coordinates of the system; they do not contain time as a variable. This means that we are confining our attention to systems where the probability of finding the particle concerned at various points in space does not vary with time. This does not mean that the particle does not move, but merely that the probability distribution associated with its movement does not vary with time. For example, an electron orbiting a hydrogen atom moves around the nucleus, but its measurable properties do not vary with time. Such systems are known as stationary states, and they include all the stable states of atoms and molecules.

1.4.5 Well-behaved Wavefunctions

Wavefunctions normally have to meet the three conditions listed below.

1. There can be only one value for ψ at any particular point. This condition is necessary to prevent there being more than one probability of finding the particle at a given location.

2. The wavefunction must vary smoothly with x so that there are no discontinuities in ψ or its first derivative, $d\psi/dx$. This arises because of the need to be able to define ψ and $d\psi/dx$ at every point. This rule does not apply to $d\psi/dx$ when the potential energy becomes infinite, as happens at the nucleus of an atom.
3. The integral of $\psi^*\psi$ over all space must be equal to one because the particle is certain to be somewhere. This requires the wavefunction to remain finite for large x.

Worked Problem 1.7

Q Which of the wavefunctions shown in Figure 1.26 are well behaved? Give reasons for your answer.

A Only the wavefunction shown in (c) is well behaved. The one shown in (a) is tending towards infinity, and therefore violates rule (3). The one shown in (b) has multiple values of ψ for a given value of x, and the one shown in (d) has discontinuities in the gradient, so that $d\psi/dx$ cannot be defined at certain points.

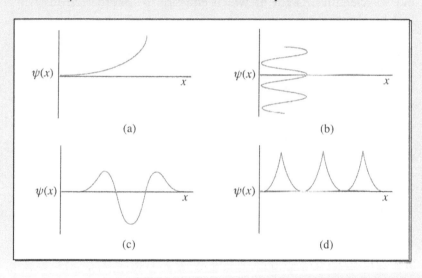

Figure 1.26 Well-behaved and unacceptable wavefunctions

Summary of Key Points

1. Electromagnetic radiation comes in small energy packets, known as photons. These have an energy equal to $h\nu$, where h is Planck's constant and ν is the frequency. Photons also have a momentum, p, which is given by the equation, $p = h/\lambda$, where λ is the wavelength.

2. Matter is also found to have both particle and wave-like properties. The wavelength of the matter wave can be obtained from the de Broglie relation, $\lambda = h/p$. This equation is similar to the one that applies to electromagnetic radiation.

3. The wavefunction for a particle moving along the x axis with constant momentum is found to be $\psi = e^{ikx}$, where k is a constant. This shows that the wavefunction can have both real and complex parts. The probability of finding the particle between x and $x + \mathrm{d}x$ is equal to $\psi^*\psi\,\mathrm{d}x$, where ψ^* is the complex conjugate of ψ. In three dimensions, the probability of finding the particle in an element of volume, $\mathrm{d}V$, is equal to $\psi^*\psi\,\mathrm{d}V$.

Problems

1.1. The work function of sodium is 2.75 eV. Calculate the threshold wavelength of light at which emission of photoelectrons will first be observed.

1.2. The threshold wavelength for photoemission of electrons from potassium is 564 nm. Calculate the maximum velocity of the photoelectrons that will be emitted when the metal is irradiated by light with a wavelength of 300 nm.

1.3. The ionization energies of electrons in the three highest occupied molecular orbitals of carbon monoxide are 14.0, 16.8 and 19.7 eV. Calculate the kinetic energies in eV of the photoelectrons that will be emitted when CO is irradiated with (a) helium radiation of wavelength 58.4 nm and (b) neon radiation of wavelength 74.2 nm.

1.4. Calculate the wavelength of a neutron beam moving with a speed of 1600 m s^{-1}. Mass of a neutron = 1.674×10^{-27} kg.

1.5. Calculate the wavelengths of: (a) a beam of electrons accelerated through a potential difference of 100 V; (b) a beam of hydrogen molecules originating from a nozzle source at a temperature of 300 K (kinetic energy of hydrogen molecules = $3.5\, k_{\mathrm{B}} T_{\mathrm{nozzle}}$, where k_{B} is Boltzmann's constant). The beams strike a close-packed nickel surface at normal incidence. If the separation between the rows of close-packed atoms is 216 pm, calculate the angle to the perpendicular at which first-order diffraction peaks will be observed for each beam.

1.6. Calculate the wavelength of a supersonic beam of argon leaving a nozzle source at a temperature of 300 K (kinetic energy of argon atoms = 2.5 $k_B T_{nozzle}$). Is diffraction likely to be observed when this beam is incident on a close-packed nickel surface?

1.7. An electron beam is accelerated through a potential difference of 50 V. For an electron in this beam, calculate (i) the momentum, (ii) the de Broglie wavelength and (iii) the magnitude of the wavevector k.

1.8. The wavefunction of a particle confined to the x axis is $\psi = e^{-x}$ for $x > 0$ and $\psi = e^{+x}$ for $x < 0$. Show that the wavefunction is normalized, and calculate the probability of finding the particle between $x = -1$ and $x = 1$.

1.9. State, giving your reasons, which of the following functions would make satisfactory wavefunctions for all values of the variable x: (i) Ne^{ax^2}; (ii) Ne^{-ax^2}; (iii) $(Ne^{-ax^2})/(3 - x)$; and (iv) Ne^{-ax}. N and a are constants.

References

1. J. H. D. Eland, *Photoelectron Spectra,* Open University Press, Milton Keynes, 1977.
2. G. Attard and C. Barnes, *Surfaces,* Oxford University Press, Oxford, 1998.
3. D. O. Hayward and A. O. Taylor, *J. Phys. C,* 1986, **19**, L309.

Further Reading

P. A. Cox, *Introduction to Quantum Theory and Atomic Structure*, Oxford University Press, Oxford, 1996, chaps. 1 & 2.
R. Eisberg and R. Resnick, *Quantum Physics of Atoms, Molecules, Solids, Nuclei and Particles*, 2nd edn., Wiley, New York, 1992, chaps. 2 & 3.
H. Haken and H. C. Wolf, *The Physics of Atoms and Quanta*, 5th edn., Springer, Berlin, 1996, chaps. 5 & 6.

2
Particle in a One-dimensional Box

Aims

By the end of this chapter you should be able to:

* Obtain the wavefunctions and energy levels for a particle-in-a-box
* Apply the particle-in-a-box model to electrons in one-dimensional semiconductor quantum wells and to π electrons in conjugated molecules

2.1   Allowed Wavefunctions and Energies

In this section we consider a particle trapped in a one-dimensional potential well with infinitely high sides, as shown in Figure 2.1. Later in the chapter this model will be applied to electrons trapped in so called "quantum wells" and also to the delocalized π electrons in conjugated molecules such as butadiene.

With an infinitely high wall there is no possibility of the particle escaping from the box, and the wavefunction outside the box must therefore be zero. It follows that the wavefunction inside the box must go to zero at the points $x = 0$ and $x = L$, otherwise there would be a discontinuity in the wavefunction at these points, and two separate values for the wavefunction would apply to the same point in space. This condition is known as a boundary condition

If an attempt is made to fit the de Broglie wavefunction, $\psi = Ae^{ikx}$, into the box, a problem arises because this function never goes to zero, and it cannot satisfy the boundary conditions. When the sine or the cosine component is equal to zero, the other component is either at a maximum or a minimum. The inability of this type of wavefunction to fit the boundary conditions is not at all surprising because it corresponds

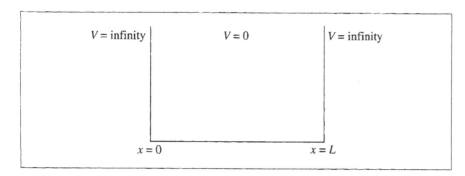

Figure 2.1 A square potential well with infinitely high sides. The potential energy, V, is zero inside the box and infinite outside

to a particle moving with constant momentum $\hbar k$, whereas a particle-in-a-box will be continually colliding with the walls and having its momentum reversed. This is illustrated in Figure 2.2. We are therefore looking for a wavefunction that represents a particle which is equally likely to be moving in either direction. This wavefunction can be obtained by a linear combination of the two exponential wavefunctions representing momentum in the +ve and –ve x directions:

Just a reminder that $\hbar$ is a shorthand way of writing $h/(2\pi)$.

$$\psi = Ae^{ikx} \pm Ae^{-ikx} \qquad (2.1)$$

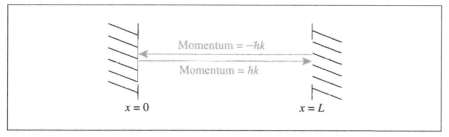

Figure 2.2 Reversal of momentum as the particle strikes the walls of the potential well

When the exponential terms are expanded, this equation becomes:

$$\psi = A[\{\cos kx + i\sin kx\} \pm \{\cos(-kx) + i\sin(-kx)\}]$$
$$= A[(\cos kx + i\sin kx) \pm (\cos kx - i\sin kx)]$$
$$= 2A\cos kx \ \text{ or } \ 2iA\sin kx \qquad (2.2)$$

The sine function satisfies the boundary conditions when $k = n\pi/L$, where n is a positive integer, known as a quantum number. When $x = 0$, $\psi = 0$ and when $x = L$, $\psi = 2iA\sin(n\pi) = 0$. Thus, the appropriate wavefunction for a particle in a box is:

$$\psi = N\sin\left(\frac{n\pi x}{L}\right) \qquad (2.3)$$

The cosine function could also be made to satisfy the boundary conditions, but only if the sides of the box were placed at $x = L/2$ and $x = 3L/2$.

with $N = 2iA$ and $n = 1, 2, 3,$ *etc.*

From equation (1.14) the kinetic energy, T, of the particle is equal to $\hbar^2 k^2/(2m)$. With the substitution $k = n\pi/L$ this becomes:

$$T = \hbar^2(n\pi)^2/(2mL^2) = n^2h^2/(8mL^2) \tag{2.4}$$

The potential energy, V, is zero and therefore the total energy, E, equals T. Thus:

$$E = \frac{n^2h^2}{8mL^2} \text{ with } n = 1, 2, 3, \textit{etc.} \tag{2.5}$$

It should be noted that $n = 0$ is not allowed because this would correspond to the wavefunction being zero everywhere, and we would have lost our particle. Thus, the particle must have a minimum energy of $h^2/8mL^2$. This is known as the zero point energy The wavefunctions are shown diagrammatically in Figure 2.3, together with the probability densities.

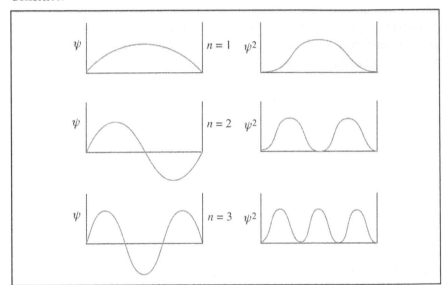

Figure 2.3 Wavefunctions and probability densities for a particle in a box

2.2   Normalization

The constant N in equation (2.3) is known as the normalization constant, and its value can be determined by using the fact that the particle is certain to be in the box. Therefore integration of the probability density, $\psi^*\psi$, over all the space available to the particle should give an answer of one, that is:

$$\int_{x=0}^{x=L} \psi * \psi \mathrm{d}x = 1 \tag{2.6}$$

With the use of the wavefunction given in equation (2.3), this becomes:

$$N^2 \int_{x=0}^{x=L} \sin^2\left(\frac{n\pi x}{L}\right) \mathrm{d}x = 1 \tag{2.7}$$

The integral can be evaluated by using the relationship $\sin^2\theta = \frac{1}{2}(1 - \cos2\theta)$. This gives:

$$N^2 \int_{x=0}^{x=L} \frac{1}{2}\left[1 - \cos\left(\frac{2n\pi x}{L}\right)\right] dx = 1 \qquad (2.8)$$

which, after integration, becomes:

$$\frac{1}{2}N^2\left[L - \left(\frac{L}{2n\pi}\right)\{\sin(2n\pi) - \sin(0)\}\right] = 1 \qquad (2.9)$$

Since $\sin(2n\pi) = 0$ and $\sin(0) = 0$, the normalization constant can be written simply as:

$$N = \sqrt{\frac{2}{L}} \qquad (2.10)$$

Hence:

$$\psi = \sqrt{\frac{2}{L}}\sin\left(\frac{n\pi x}{L}\right), \; n = 1, 2, 3, \; etc. \qquad (2.11)$$

2.3 Probability Distributions of the Wavefunctions

It should be noted that the probability densities shown diagrammatically in Figure 2.3 are not what one would expect on the basis of classical mechanics, and they raise a number of conceptual problems. A particle that is moving with constant speed, and suffering a reversal of direction every time it collides with the walls, would be expected to spend equal lengths of time in all parts of the box. It follows that the probability of finding the particle at various points inside the box at some arbitrary moment in time should be constant. In fact, when the wave nature of the particle is taken into account, we find that the particle in its ground state ($n = 1$) is most likely to be found at the centre of the box.

Conceptual problems become even greater when we consider the $n = 2$ state, where the probability density drops to zero at the centre of the box. How can the particle get from one side of the box to the other if there is no chance of finding it at the centre? There is no simple answer to this question, but the problem will be partially resolved when we consider the Heisenberg Uncertainty Principle in Chapter 3. The wavefunctions that we have considered so far relate to particles that have specific energies but indeterminate positions. As we shall see later, any attempt to locate the particle interferes with the system, and changes its wavefunction.

2.4   Semiconductor Quantum Wells

These are sandwich structures made from semiconducting materials and they have many potential applications in modern electronic devices. An example is illustrated in Figure 2.4, where the base material is the III/V semiconductor gallium arsenide (GaAs). A thin layer of pure GaAs has been created between two layers of aluminium gallium arsenide, a material in which some of the gallium has been replaced with aluminium to give the formula $Al_xGa_{1-x}As$. Nearly free electrons, known as conduction electrons, can exist in these semiconductors, and they have a much higher potential energy in $Al_xGa_{1-x}As$ than they do in pure GaAs. Thus, conduction electrons in the pure GaAs become trapped between two potential walls and behave like particles in a one-dimensional box when moving in the x direction, although they have complete freedom of movement in the y and z directions. Unlike the idealized potential well considered earlier, which had walls that were infinitely high, this one has finite walls. This affects the wavefunctions and energies to a small extent, but the equations derived earlier will still give reasonably accurate values for the energies.

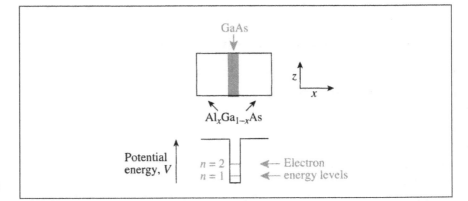

Figure 2.4 The variation of potential energy across a gallium arsenide one-dimensional potential well, and the associated electron energy levels

The existence of discrete energy levels is confirmed by the observation of selective absorption of laser light at certain frequencies which correspond to the transition of an electron from one energy level to another. The GaAs/$Al_xGa_{1-x}As$ combination mentioned above forms the basis of the semiconductor laser used in compact disc players. Equation (2.5) can be used to calculate the energies, with one important proviso: the conduction electrons behave as though they have a much smaller mass than ordinary electrons. This mass is known as the effective mass, m^*; for gallium arsenide, $m^* = 0.067m_e$.

Worked Problem 2.1

Q What is the wavelength of radiation emitted when an electron in a gallium arsenide potential well of width 10.0 nm drops from the first excited state to the ground state?

A The first excited state corresponds to $n = 2$ and the ground state to $n = 1$. The difference in energies is obtained as follows:

$$\Delta E = \frac{(2^2 - 1^2)h^2}{8m*L^2} = \frac{3 \times (6.626 \times 10^{-34} \text{ J s})^2}{8 \times 0.067 \times (9.109 \times 10^{-31} \text{ kg}) \times (10 \times 10^{-9} \text{ m})^2}$$
$$= 2.697 \times 10^{-20} \text{ J}$$

The wavelength of the radiation can be obtained by using the equation $\Delta E = h\nu = hc/\lambda$.
Thus:

$$\lambda = \frac{(6.626 \times 10^{-34} \text{ J s}) \times (2.998 \times 10^8 \text{ m s}^{-1})}{2.697 \times 10^{-20} \text{ J}} = 7.36_5 \times 10^{-6} \text{ m}$$

This wavelength lies in the infrared region of the spectrum.

2.5   π Electrons in Conjugated Molecules

Butadiene, $CH_2{=}CH{-}CH{=}CH_2$, will be used to illustrate the way in which the particle-in-a-box wavefunctions can be applied to conjugated molecules, a term applied to hydrocarbon molecules with alternate single and double carbon–carbon bonds. To understand the nature of the bonding in such molecules it is necessary to anticipate the shapes of the atomic orbitals, a subject that will be discussed more fully in Chapter 6.

Figure 2.5 shows the way in which the 2s, $2p_x$ and $2p_y$ atomic orbitals on each carbon atom combine to give three equivalent orbitals, known as hybrid orbitals. The overlap of these hybrid orbitals in the xy plane produces so called "σ bonding", and results in the molecular shape shown in Figure 2.6. The $2p_z$ orbitals on the carbon atoms are directed at right angles to the xy plane, as shown in Figure 2.7, and they are not involved in the σ bonding. Instead, they overlap to give what are termed "π orbitals". It can be seen that the $2p_z$ orbitals on the second and third carbon atoms can interact equally well with similar orbitals on either side. This results in delocalized orbitals that extend over the whole molecule. Electrons in these π orbitals can move from one end of the molecule to the other, and can be thought of as particles in a box.

An electron attached to an atom can have a variety of wavefunctions, and these are known as atomic orbitals. Spherically symmetric wavefunctions are called s orbitals. There are also dumb-bell shaped orbitals aligned along the three Cartesian axes, and these are known as p_x, p_y and p_z orbitals.

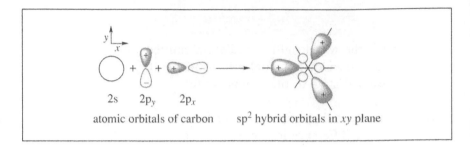

Figure 2.5 Formation of sp² hybrid orbitals in the *xy* plane of a carbon atom

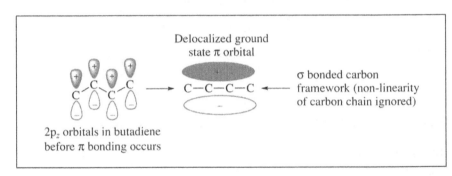

Figure 2.6 The overlap of carbon sp² hybrid orbitals in the *xy* plane of the butadiene molecule

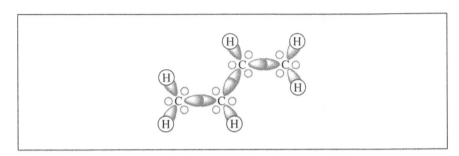

Figure 2.7 Formation of a delocalized π orbital by the overlap of the carbon 2p$_z$ orbitals in butadiene

A more sophisticated method of treating the π electrons in conjugated molecules, known as the Hückel approximation, will be fully discussed in Chapter 8.

To simplify the problem further we will assume that the carbon framework is linear and that the potential energy of the π electrons remains constant as they move along the molecule. The second assumption may seem to be a poor approximation because of the strong coulombic attraction that exists between the negatively charged electron and the positively charged carbon nuclei. This should result in a substantial fall in the potential energy of the electron as it passes close to a nucleus. However, this effect will not be large for π electrons because, as can be seen in Figure 2.7, the electron density is concentrated at points above and below the internuclear axis and there is actually a node along the axis itself. Thus, the π electrons never come very close to the carbon nuclei during their movement along the molecule. The model and actual potentials are compared in Figure 2.8. The actual value of V to be used is arbitrary, provided it remains constant, and we shall put $V = 0$.

Two electrons can occupy each state, with their spins paired. Since

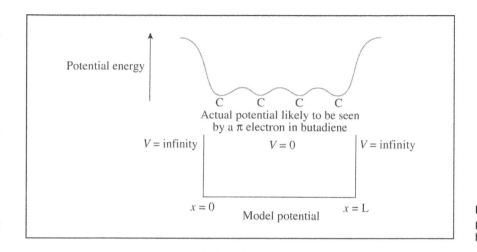

Figure 2.8 Model and actual potentials seen by a π electron in butadiene

butadiene has four π electrons, the two lowest energy states will be fully occupied, as shown in Figure 2.9. Experimentally it is found that butadiene absorbs electromagnetic radiation with a wavelength, λ, equal to 217 nm. The energy of the photon is used to excite a π electron from the $n = 2$ to the $n = 3$ state, and the wavelength of the radiation can be calculated from the equation $E_3 - E_2 = h\nu = hc/\lambda$. This gives a value of 220 nm, which is in surprisingly good agreement with the experimental value.

The spin of the electron will be discussed in Chapter 5. Electrons occupying the same spatial orbital must have opposite spins and this restricts the number of electrons that can occupy the state to two.

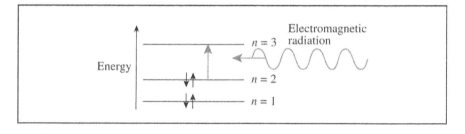

Figure 2.9 Occupation of energy levels in butadiene. Absorption of electromagnetic radiation causes promotion of an electron from the $n = 2$ to the $n = 3$ level

The π electron probability densities for the first three states are shown in Figure 2.10. The wavefunctions must extend beyond the carbon atom framework, otherwise there would always be a node on the two end carbon atoms, which would indicate zero π electron density there. If the box is extended by half a bond length at each end of the molecule, the length of the box becomes equal to four carbon–carbon bond lengths. For $n = 2$ the π electron density is concentrated between carbon atoms 1 and 2, and 3 and 4, making these approximate to double bonds. There is a node between the two central carbon atoms, so this must be a simple σ bond. The $n = 2$ state therefore corresponds to the structure normally drawn for butadiene. The excited state, with $n = 3$, has zero π electron probability density between atoms 1 and 2, and also between atoms

3 and 4, but a maximum probability density between the middle two atoms. It therefore corresponds very roughly to the structure $CH_2–CH=CH–CH_2$, which has unsatisfied valence on the end carbon atoms. This makes the excited state particularly reactive.

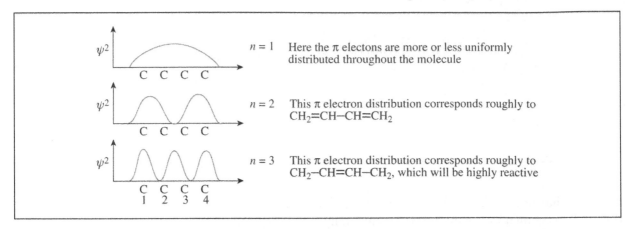

$n = 1$ Here the π electons are more or less uniformly distributed throughout the molecule

$n = 2$ This π electron distribution corresponds roughly to $CH_2=CH–CH=CH_2$

$n = 3$ This π electron distribution corresponds roughly to $CH_2–CH=CH–CH_2$, which will be highly reactive

Figure 2.10 π Electron probability densities in butadiene

Worked Problem 2.2

Q Calculate the wavelength of the radiation that will be absorbed in promoting an electron from the highest occupied molecular orbital (HOMO) to the lowest unoccupied molecular orbital (LUMO) in butadiene.

A First, an estimate has to be made of the length of the potential well, L, in which the π electrons move. It seems reasonable to take the mean C–C distance to be equal to the average of the carbon–carbon single and double bond lengths found in a variety of non-conjugated compounds. With $d(C–C) = 154$ pm and $d(C=C) = 135$ pm, this leads to a value of 144.5 pm. The total length of the box is then taken to be equal to four carbon–carbon bond lengths, giving $L = 578$ pm.

The highest occupied π electron state has $n = 2$, and the lowest unoccupied state has $n = 3$. The energy required to promote an electron from one state to the other is given by the equation:

$$E_3 - E_2 = \frac{(3^2 - 2^2)h^2}{8mL^2} = \frac{5 \times (6.626 \times 10^{-34} \text{ J s})^2}{8 \times (9.109 \times 10^{-31} \text{ kg}) \times (578 \times 10^{-12} \text{ m})^2}$$

$$= 9.017 \times 10^{-19} \text{ J}$$

Hence:

$$\lambda = \frac{hc}{E} = \frac{(6.626 \times 10^{-34} \text{ J s}) \times (2.998 \times 10^{8} \text{ m s}^{-1})}{9.017 \times 10^{-19} \text{ J}} = 2.203 \times 10^{-7} \text{ m} = 220 \text{ nm}$$

Summary of Key Points

1. The wavefunction for a particle in a one-dimensional box of length L was found to be:

$$\psi = \sqrt{\frac{2}{L}} \sin\left(\frac{n\pi x}{L}\right)$$

where n is a quantum number that can equal 1, 2, 3, *etc*. The energy of an electron in such a box is quantized, with energies given by the equation:

$$E = \frac{n^2 h^2}{8mL^2}$$

Here, m is the mass of the electron.

2. The particle-in-a-box wavefunction can be applied to electrons in gallium arsenide quantum wells provided the electron is given an effective mass which is much less than its true mass.

3. The particle-in-a-box wavefunction can also be applied to π electrons in conjugated hydrocarbon molecules such as butadiene. The wavelength of radiation that is absorbed when a π electron is promoted from the highest occupied energy level (HOMO) to the lowest unoccupied level (LUMO) has been calculated, and found to be reasonably close to the experimental value.

Problems

2.1. For a one-dimensional gallium arsenide quantum well of width 21 nm, calculate the difference in energies between the $n = 2$ and $n = 3$ states for travel of conduction electrons across the width of the well. Compare your answer with the experimentally determined value of approximately 0.05 eV.

2.2. Calculate the energy separation between the $n = 1$ and $n = 2$ levels of a nitrogen molecule, confined in a one-dimensional box of length 1 cm. Find the value of n that corresponds to the average thermal energy of a nitrogen molecule at a temperature of 300 K. The average thermal energy $= 1/2k_B T$ for movement in one dimension, where k_B is the Boltzmann constant. Comment on the value for n obtained.

2.3. Calculate the wavelength of light that will be absorbed when a π electron in hexa-1,3,5-triene, $CH_2=CH-CH=CH-CH=CH_2$, is promoted from the highest occupied level to the lowest unoccupied level. The average C–C bond length in hexatriene can be taken to be 144 pm. Compare your answer with the experimentally observed wavelength of 258 nm.

2.4. β-Carotene contains a conjugated chain of 22 carbon atoms with an average internuclear separation of 144 pm. The π electrons in this conjugated chain are delocalized and, to a rough approximation, they can be considered to be particles in a one-dimensional box of length 3.17 nm (that is, 22×144 pm). Each carbon atom contributes one π electron. Calculate: (a) the difference in energy between the highest occupied π electron energy level and the lowest unoccupied level; (b) the wavelength of radiation that would be absorbed in producing a transition between these two states; (c) the average number of π electrons likely to be found in the region between carbon atoms 11 and 12 in the ground state of the molecule; and hence (d) the bond order. The answer to part (c) can be found by summing the probabilities of each electron being between carbon atoms 11 and 12. The wavefunctions can be taken to be roughly constant over the small distance involved; therefore no integration is necessary. Lengthy mathematics are not required for this calculation, just a little bit of *nous*!

2.5. A particle is in a one-dimensional box that extends from $x = 0$ to $x = L$. Obtain a general expression for the probability of finding the particle between $x = 0$ and $x = L/4$ in terms of the quantum number n. Hence calculate the probabilities for $n = 1, 2, 3$ and 4. Sketch the way in which the probability varies as n tends towards infinity. The trigonometric formula $\sin^2\theta = \frac{1}{2}(1 - \cos 2\theta)$ may be found useful in the integration of $\sin^2(n\pi x/L)$.

Further Reading

R. L. Flurry, Jr., *Quantum Chemistry: an Introduction,* Prentice-Hall, Englewood Cliffs, NJ, 1983, chap. 2.

F. L. Pilar, *Elementary Quantum Chemistry*, 2nd edn., McGraw-Hill, New York, 1990, chap. 3.

3

Uncertainty Arising from the Wave Nature of Matter

Aims

By the end of this chapter you should be able to:

*   Understand how the uncertainties in position and momentum of a particle arise from the wave nature of matter
*   Use the Uncertainty Principle to calculate minimum values for these uncertainties

3.1   Uncertainty in the Diffraction of Particles

In classical mechanics the trajectory followed by a particle is fully determined by the starting conditions, and the position of the particle as a function of time can be specified to any desired accuracy. This implies that precise values can be given to both the position and momentum of the particle at any instant. The elastic scattering of heavy atoms from surfaces (*e.g.* argon and krypton) comes within this category and results in "rainbow scattering" under appropriate conditions, as shown in Figure 3.1. Here the point of impact is localized and the momentum after impact can be determined from the surface corrugation. Although heavy atoms have wave-like properties, the wavelength is very much shorter than the interatomic distances found in solids, and diffraction is not observed.

When electrons or much lighter atoms are used, however, it becomes impossible to give precise trajectories because the wave-like properties of these particles control the scattering. Take electron diffraction, for example. If a beam containing millions of electrons is fired at a nickel target, we know that the diffraction pattern shown earlier will be obtained with six first-order peaks of equal intensity, but it is not possible to know in advance to which spot any particular electron will go. Therefore the trajectory followed by a single electron cannot be

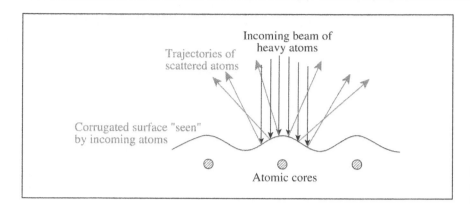

Figure 3.1 Rainbow scattering of heavy atoms from an ordered surface

determined from the starting conditions (see Figure 3.2). It follows that the deterministic nature of newtonian mechanics is destroyed and an unavoidable uncertainty is built into the detailed outcome of any experiment.

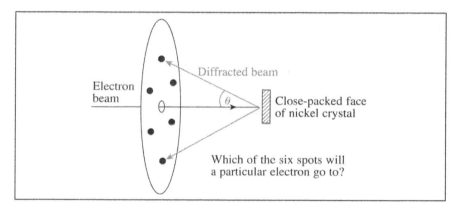

Figure 3.2 Uncertainty in low-energy electron diffraction from a nickel surface

Diffraction patterns can be built up gradually by observing the arrival of single electrons at a fluorescent screen, the time between arrivals being sufficiently long to prevent any interaction between different electrons. At first, the arrival of the electrons seems to be random, but as more electrons arrive a diffraction pattern begins to emerge. Thus, although the behaviour of a single electron is unpredictable, the collective behaviour of a very large number of electrons can be determined to a fairly high accuracy.

The phenomenon of diffraction introduces a certain "fuzziness" or uncertainty into the trajectories followed by the diffracting particles. Figure 3.3 shows the situation at the moment of collision of a diffracting particle (electron, light atom or molecule) with a surface. The first uncertainty arises in the position of impact of the particle with the surface. Diffraction can occur only if the interaction with the surface extends over more than one lattice spacing, which means that the uncertainty in the position of the particle at the moment of impact, Δx, must be equal to, or greater than, d.

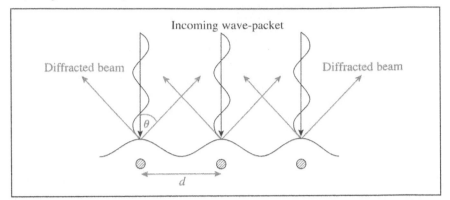

Figure 3.3 Uncertainty in position and momentum during diffraction at a surface

This can produce some surprising results. For instance, in the diffraction of a supersonic helium beam from a sodium-covered silicon surface, shown in Figure 3.4, a periodic lattice spacing of about 1.5 nm is detected between rows of sodium atoms.[1] This distance is much greater than the collision diameter obtained for helium from the kinetic theory of gases, which is about 0.5 nm. In fact, the discrepancy is even greater than this because a reasonably sharp diffraction pattern will require contributions from at least three rows of sodium atoms. This increases the distance over which a helium atom can interact with the surface to about 3 nm. It must be concluded that the behaviour of a helium atom is far removed from the commonly held picture of a small billiard ball bouncing off a rough surface. The entity undergoing diffraction has to be treated as a wave packet, which extends beyond the normal dimensions of the atom.

There is also an uncertainty in the momentum of the particle as it

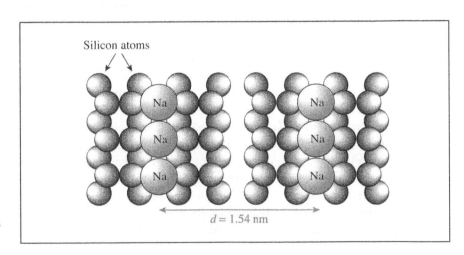

Figure 3.4 Lattice spacing for a sodium-covered silicon surface

leaves the surface because, as we have already seen, it is not possible to know in advance to which diffraction spot the particle will go. In Figure 3.3, the particle is equally likely to be diffracted to the left or to the right. From the de Broglie relation, the magnitude of the momentum will be h/λ, and the component parallel to the surface will therefore be equal to $\pm h\sin\theta/\lambda$, the sign depending on whether the particle is diffracted to the left or the right. The uncertainty in the component of momentum parallel to the surface, Δp_x, will therefore also be equal to $h\sin\theta/\lambda$. Use of the diffraction relation $n\lambda = d\sin\theta$, with $n = 1$, leads to $\Delta p_x = h/d$. This is only an approximate formula because the other four first-order diffraction peaks have been ignored, as well as higher-order peaks, but it gives a rough measure of the overall uncertainty in the momentum. When this uncertainty is combined with the uncertainty in the point of impact, we obtain the relation:

$$\Delta x \Delta p_x \approx h \qquad (3.1)$$

It is significant that the particular properties of the system have cancelled out, and we are left simply with a universal constant. It is also worth noting that the "uncertainty" or "fuzziness" that we have discussed does not arise from any defect in our knowledge of the system, or the method of measurement, but from the inherent nature of the diffraction process.

3.2   Diffraction of Electrons through Double Slits

The uneasy combination of particle and wave-like properties can lead to some strange results. Perhaps the most famous one occurs when electrons (or any other light particles) are diffracted by a pair of slits, as shown in Figure 3.5. Waves emerging from the two slits interfere with one another to create, on a fluorescent screen, the diffraction pattern shown. This consists of alternating bright and dark fringes where the waves from the two slits are either in or out of phase with one another.

A difficulty arises when one considers what happens to a single electron as it passes through the double slit system. Each electron will give rise to a spot on the screen and the complete diffraction pattern can be built up by an accumulation of these spots, without any interaction occurring between the different electrons. Let us assume for the moment that an electron, being a particle, can pass through only one of the two slits, which we will take to be slit A. The probability of it striking various points on the screen will depend upon whether slit B is open or closed. If it is closed the probability of the electron striking various points on the screen will be determined by the diffraction pattern shown in the lower part of the diagram, whereas the upper diffraction pattern will apply if slit B is open. How does the electron "know" whether slit B is

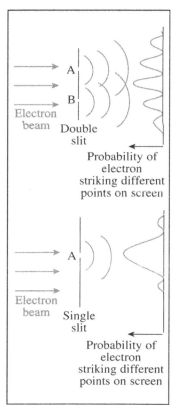

Figure 3.5 Electron diffraction at double and single slits

By this stage you may feel that quantum mechanics is not for you. If so, take heart from the Nobel Prize winner Richard Feynman, who once made the remark "I think I can safely say that nobody understands quantum mechanics".

open or closed? Could it be that in some sense the electron actually goes through both slits? The fact that the diffraction pattern can be built up by separate, non-interacting electrons seems to support the idea that each electron passes through both slits and interferes with itself. Or does this seem too absurd to be taken seriously?

3.3   Uncertainty with Particle-in-a-box

See equation (2.5).

A similar uncertainty arises with the particle in a box. Although the energy of the particle is known, there is an uncertainty in the momentum because it is not possible to know at any given instant whether the particle is moving from left to right or right to left. Combining the equations for energy and momentum, $E = n^2h^2/8mL^2$ and $p^2 = 2mE$, it can be seen that $p = \pm nh/2L$. There is, therefore, an uncertainty in the momentum, Δp_x, equal to $nh/2L$. There is also an uncertainty in position because we do not know where the particle is inside the box. If this uncertainty is put equal to half the length of the box, $L/2$, we find that, for the ground state:

$$\Delta x \Delta p_x \approx \left(\frac{h}{2L} \right)\left(\frac{L}{2} \right) = \frac{h}{4} \qquad (3.2)$$

Once again, the product of the two uncertainties is proportional to Planck's constant, and does not involve any parameters specific to the system. There is no way to avoid this basic uncertainty. Making the box smaller will reduce the uncertainty in x, but only at the expense of increasing the uncertainty in p_x. It is also significant that the particle in a box is not allowed to have zero energy, corresponding to $n = 0$. Zero energy would mean that there was no momentum, and therefore Δp_x would be equal to zero. We shall see in the next section that this would violate the Uncertainty Principle.

A more sophisticated calculation[2] uses standard deviations from the mean to obtain the uncertainties in x and p_x. When this is done, Δp_x is still found to be equal to $h/(2L)$ but Δx is now much smaller and equal to $0.57L/\pi$. Thus:

$$\Delta x \Delta p_x = 0.57h / (2\pi) = 0.57\hbar \qquad (3.3)$$

3.4   Statement of the Heisenberg Uncertainty Principle

Because of his work on the Uncertainty Principle, Heisenberg was awarded the Nobel Prize for Physics in 1934.

Heisenberg, by considering a hypothetical experiment in which the position and momentum of an electron were deduced from observations made in an optical microscope, came to the general conclusion that the uncertainty in linear momentum of a particle, Δp_x, multiplied by the uncertainty in its position, Δx, could never be less than $h/4\pi$. This can be written formally as:

$$\Delta x \Delta p_x \geq h/(4\pi) = \hbar/2 \qquad (3.4)$$

Thus, it is not possible to know simultaneously both the precise position and the momentum of a microscopic particle, such as an electron or atom.

Worked Problem 3.1

Q For a particle moving freely along the x axis, show that the Heisenberg uncertainty principle can be written in the alternative form:

$$\Delta \lambda \Delta x \geq \lambda^2/(4\pi)$$

where Δx is the uncertainty in position of the particle and $\Delta \lambda$ is the simultaneous uncertainty in the de Broglie wavelength.

A Differentiation of the de Broglie relation, $p = h/\lambda$, gives $dp/d\lambda = -h/\lambda^2$. The uncertainty in momentum, Δp_x, can be equated with dp and the uncertainty in wavelength, $\Delta \lambda$, with $-d\lambda$ (the uncertainties must always be positive). Thus:

$$\frac{\Delta p_x}{\Delta \lambda} = \frac{h}{\lambda^2} \text{ and } \Delta \lambda \Delta x = \frac{\lambda^2}{h} \Delta p_x \Delta x$$

Combining this with equation (3.4) we obtain:

$$\Delta \lambda \Delta x \geq \frac{\lambda^2}{h} \left(\frac{h}{4\pi} \right) = \frac{\lambda^2}{4\pi}$$

which is the desired result.

3.5 Application to an Electron Beam

The electrons in a beam move with constant momentum along a specific direction, taken to be the x direction (see Figure 3.6). We can see how the Uncertainty Principle applies by considering the wavefunction of an electron in the beam:

$$\psi = Ae^{ikx} \tag{3.5}$$

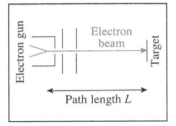

Figure 3.6 An electron beam in which the electrons have a well-defined momentum, but indeterminate position

The probability, $P(x)\mathrm{d}x$, of finding the electron between x and $x + \mathrm{d}x$ is given by the equation:

$$P(x)\mathrm{d}x = \psi\psi^*\mathrm{d}x = A^2 e^{-ikx}e^{ikx}\mathrm{d}x = A^2\mathrm{d}x \tag{3.6}$$

This shows that the electron is equally likely to be anywhere along the x axis. In other words, the position of the electron is unknown, and the uncertainty in its position, Δx, is equal to the path length L, which is very large compared to the wavelength of the electron. On the other hand, the momentum of the electron is known fairly precisely. The value is given by the equation:

$$p = h/\lambda = h/(2\pi/k) = hk/2\pi = \hbar k \tag{3.7}$$

Since the value of k can be determined to quite a high accuracy, the uncertainty in the momentum is very small. This is the normal situation with an electron beam, where there is a constant flux of electrons of known energy, but the actual position of individual electrons within the beam is unknown and is of no practical interest. For this situation, Δx is very large and Δp is very small, and the product of the two must be greater than $\hbar/2$.

Worked Problem 3.2

Q A beam of electrons travels with a velocity of 1000 ± 0.01 m s^{-1}. In terms of the uncertainty principle, how precisely can the position of an electron in the beam be defined at any instant?

A The uncertainty in momentum is given by the equation:

$$\Delta p_x = \left(0.01 \text{ m s}^{-1}\right) \times \left(9.11 \times 10^{-31} \text{ kg}\right) = 9.11 \times 10^{-33} \text{ kg m s}^{-1}$$

Hence, the uncertainty in the position of the electron is given by the expression:

$$\Delta x \geq \frac{\hbar}{2 \times 9.11 \times 10^{-33} \text{ kg m s}^{-1}} = 0.0058 \text{ m} = 0.58 \text{ cm}$$

3.6   The Wavefunction of a Localized Electron

In this section we consider the situation in which the position of an electron has been determined very precisely so that Δx approaches zero. What will the wavefunction of such an electron look like? Bearing in mind that the probability of finding the electron between x and $x + \mathrm{d}x$ is equal to $\psi\psi^*\mathrm{d}x$, we can expect the wavefunction to consist of a delta function, as shown in Figure 3.7.

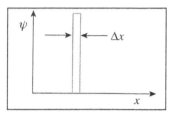

Figure 3.7 The wavefunction of a particle located between x and $x + \Delta x$

This type of wavefunction presents us with an immediate problem: how can we derive an electron energy or momentum from it? The answer is to regard the delta function as the sum, or superposition, of an infinite number of sinusoidal waves with different wavelengths. This is illustrated in Figure 3.8. When these waves are all in phase at one particular point, and nowhere else, they will generate a delta function.

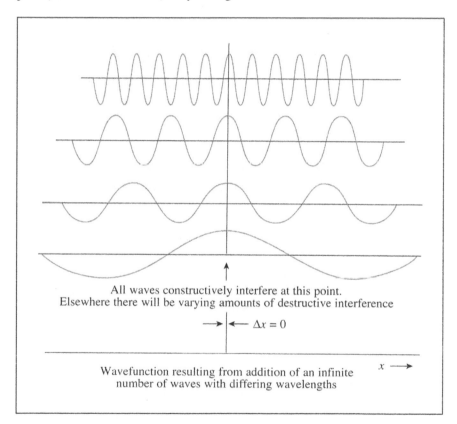

All waves constructively interfere at this point. Elsewhere there will be varying amounts of destructive interference

$\Delta x = 0$

Wavefunction resulting from addition of an infinite number of waves with differing wavelengths

Figure 3.8 The superposition of an infinite number of waves which are in phase at only one point

Since the delta function is made up of an infinite number of sinusoidal waves with differing wavelengths, and therefore differing k values, there is an infinite uncertainty in the momentum of the particle, with each sinusoidal wave representing a different momentum according to the formula $p = \hbar k$.

Conclusion: the position of the particle is known, but there is an infinite uncertainty as to its momentum.

Summary of Key Points

The Heisenberg Uncertainty Principle states that, for a particle moving along the x axis, the uncertainty in its position, Δx, multiplied by the uncertainty in its momentum, Δp_x, can never be less than $\hbar/2$, where $\hbar$ is a shorthand notation for $h/2\pi$. This can be written as:

$$\Delta x \Delta p_x \geq \hbar / 2$$

This uncertainty arises from the wave nature of matter.

Problems

3.1. The position of an electron moving in one dimension is determined to an accuracy of ±0.01 nm by reflection of a photon. Calculate the simultaneous uncertainty in the velocity of the electron at the moment of localization.

3.2. If the uncertainty in locating a particle moving in one dimension is equal to its de Broglie wavelength, calculate the fractional uncertainty in its velocity, $\Delta v/v$.

3.3. To a rough approximation, an electron in the ground state of a hydrogen atom can be considered to be like a particle in a one-dimensional box with a length equal to twice the Bohr radius, a_0. It can be shown that the average kinetic energy of the electron is equal to $\hbar^2/(2m_e a_0^2)$, where m_e is the mass of the electron. Estimate the uncertainties in position Δx and momentum Δp_x, and show that $\Delta x \Delta p_x \approx \hbar$.

Hint: although the average kinetic energy is known, the direction of motion at any instant is not, and the electron is equally likely to be travelling towards or away from the nucleus.

3.4. The Heisenberg uncertainty principle can be applied to radiation in the form derived in Worked Problem 3.1. If the wavelength of a photon can be measured to one part in 10^7, calculate the uncertainty in the position of (a) an X-ray photon with a wavelength of 0.5 nm, and (b) a visible light photon with a wavelength of 500 nm.

References

1. T. Bush, PhD thesis, University of London, 1995.
2. D. A. McQuarrie, *Quantum Chemistry*, University Science Books/Oxford University Press, Sausalito, California/Oxford, 1983, pp. 95, 96.

Further Reading

R. Eisberg and R. Resnick, *Quantum Physics of Atoms, Molecules, Solids, Nuclei and Particles*, 2nd edn., Wiley, New York, 1985.

4

The One-dimensional Schrödinger Wave Equation and Some of its Applications

Aims

By the end of this chapter you should be able to:

- Use the one-dimensional Schrödinger equation to calculate the energies of particles with simple wavefunctions
- Use the harmonic oscillator model to calculate the zero point energies and force constants of diatomic molecules
- Calculate the tunnelling probabilities of particles passing through rectangular potential energy barriers
- Understand the operating principles of the scanning tunnelling microscope

4.1 The One-dimensional Schrödinger Equation

The simple de Broglie relation, $\lambda = h/p$, proved to be adequate to account for the properties of a particle moving with constant momentum in one dimension, but it cannot be applied to more complex systems where the momentum of the particle varies with position, or there is more than one variable. The equation which describes the wave behaviour of such systems is the celebrated Schrödinger wave equation, developed by Erwin Schrödinger in 1926. For one-dimensional systems with constant potential energy, this equation produces exactly the same results as those obtained from application of the de Broglie relation.

Because the Schrödinger wave equation is an expression of some of the most basic principles of quantum mechanics, it cannot be derived from a more fundamental equation. However, it is possible to arrive at the equation by considering a general wave equation and applying the de Broglie relation to it. This is done below.

All the wavefunctions that have been discussed so far (e^{ikx}, $\sin kx$ and $\cos kx$) are solutions of the differential equation:

$$\frac{d^2\psi}{dx^2} = -k^2\psi \qquad (4.1)$$

When differentiated twice, all these functions return to their original form, multiplied by a constant.

It is fairly easy to show that this equation will always generate waves, and Figure 4.1 illustrates this. $d^2\psi/dx^2$ is a measure of the curvature of the wavefunction (rate of change of gradient with x) and it always has the opposite sign from the wavefunction. Thus, when ψ is positive, $d^2\psi/dx^2$ is negative, and the wavefunction bends downwards towards the x axis. Once ψ becomes negative, the sign of $d^2\psi/dx^2$ changes and the wavefunction starts to bend upwards towards the x axis. In this way a wave is generated.

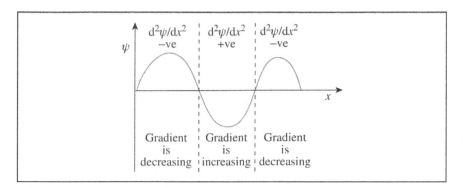

Figure 4.1 Plot of an oscillating wavefunction, showing how $d^2\psi/dx^2$ always has the opposite sign to ψ

From equation (1.14) we know that the kinetic energy of the particle, T, is equal to $(\hbar k)^2/(2m)$. Thus, we can write equation (4.1) in the form:

$$\frac{d^2\psi}{dx^2} = -\frac{2mT\psi}{\hbar^2} \qquad (4.2)$$

After rearrangement this becomes:

$$-\frac{\hbar^2}{2m}\left(\frac{d^2\psi}{dx^2}\right) = T\psi \qquad (4.3)$$

This is the wave equation for a particle moving with constant kinetic energy along the x coordinate. There is a direct relationship between the kinetic energy and the curvature of the wavefunction, which is proportional to $d^2\psi/dx^2$.

In many situations the kinetic energy of the particle is not constant because the potential energy, V, varies with the position, x. The quantity that remains constant is the total energy, E. Since $T = E - V$, we can rewrite the wave equation as:

$$-\frac{\hbar^2}{2m}\left(\frac{\mathrm{d}^2\psi}{\mathrm{d}x^2}\right) = (E - V)\psi \qquad (4.4)$$

This is the Schrödinger wave equation in one dimension. We will leave consideration of the three-dimensional Schrödinger equation until Chapter 5.

Worked Problem 4.1

Q Show that the particle-in-a-box wavefunction:

$$\psi = N\sin\left(\frac{n\pi x}{L}\right)$$

is a solution of the one-dimensional Schrödinger equation with $V = 0$, and hence calculate the allowed energies of the particle.

A For the particle-in-a-box wavefunction:

$$\frac{\mathrm{d}^2\psi}{\mathrm{d}x^2} = -N\left(\frac{n\pi}{L}\right)^2\sin\left(\frac{n\pi x}{L}\right) = -\left(\frac{n\pi}{L}\right)^2\psi$$

Clearly, ψ is a solution of equation (4.4) with:

$$E = \left(\frac{\hbar^2}{2m}\right)\left(\frac{n\pi}{L}\right)^2 = \frac{\left(n\pi\hbar\right)^2}{2mL^2} = \frac{n^2h^2}{8mL^2}$$

4.2   The One-dimensional Harmonic Oscillator

In Chapter 2 we considered a particle confined to a square potential well with infinitely high sides. In that model there was no force acting on the moving particle until it reached one of the walls, where it underwent an elastic reflection. In this chapter a different type of one-dimensional potential well will be considered in which the particle experiences a force which is proportional to its displacement from the midpoint of the well. This model is particularly suited to an examination of the vibrations of molecules.

4.2.1 Classical Treatment of a Vibrating Diatomic Molecule

The atoms of a diatomic molecule can be regarded as two particles joined together by a spring of unextended length r_0, as shown in Figure 4.2. If the spring is extended to a distance $r_0 + x$ and then let go, the molecule will begin to vibrate as a result of the action of the restoring force, which, to a first approximation, is proportional to the value of x. This is Hooke's Law , and it can be written mathematically as:

$$\text{force} = -kx \tag{4.5}$$

where k is known as the force constant of the bond. The minus sign is required because the force acts in the opposite direction to x. The work dW required to extend the bond by a distance dx is given by the equation:

$$dW = -(\text{force})dx = kx dx \tag{4.6}$$

Here, again, a minus sign is required because the work done is positive when the movement is in the opposite direction to the action of the force. The potential energy of the diatomic molecule, V, is equal to the total work done in extending the spring and is therefore given by the equation:

$$V = \int_0^x dW = \int_0^x kx\, dx = \tfrac{1}{2} kx^2 \tag{4.7}$$

where the potential energy is taken to be zero when $x = 0$.

It can be shown[1] that the vibration of a diatomic molecule about its centre of mass is mathematically equivalent to the oscillations of a single particle of mass μ [$= m_A m_B/(m_A + m_B)$] about a fixed point; see Figure 4.3. Application of Newton's second law of motion (*force = mass × acceleration*) to the vibration leads to the equation:

$$-kx = \mu \frac{d^2 x}{dt^2} \tag{4.8}$$

where t represents the time. A solution of this equation is:

$$x = A \sin\left\{ \left(\frac{k}{\mu}\right)^{1/2} t \right\} \tag{4.9}$$

where A is the maximum value of x reached during the oscillation. It can easily be checked that this is a correct solution of equation (4.8) by twice differentiating x with respect to time. The sine function, which describes an oscillation, is shown in Figure 4.4. When the time is

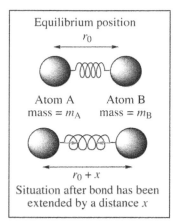

Equilibrium position
r_0

Atom A   Atom B
mass = m_A mass = m_B

$r_0 + x$

Situation after bond has been extended by a distance x

Figure 4.2 Restoring force resulting from extension of the bond length of a diatomic molecule

The hypothetical mass μ is known as the **reduced mass** of the system.

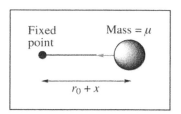

Fixed point   Mass = μ

$r_0 + x$

Figure 4.3 The oscillation of the reduced mass about a fixed point

increased by an amount τ, the time for one complete vibration, the angle of the sine function must increase by 2π. Thus:

$$\left(\frac{k}{\mu}\right)^{1/2} \tau = 2\pi \tag{4.10}$$

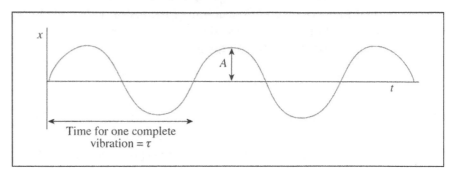

Figure 4.4 The variation of displacement with time for a harmonic oscillator

The fundamental frequency of the vibration, ω_0, is equal to $1/\tau$. After rearrangement, equation (4.10) becomes:

$$\omega_0 = \frac{1}{\tau} = \frac{1}{2\pi}\left(\frac{k}{\mu}\right)^{1/2} \tag{4.11}$$

4.2.2 Quantum Mechanical Treatment of a Vibrating Diatomic Molecule

With $V = \frac{1}{2}kx^2$, the one-dimensional Schrödinger equation for this system becomes:

$$-\frac{\hbar^2}{2\mu}\frac{d^2\psi}{dx^2} = \left(E - \frac{1}{2}kx^2\right)\psi \tag{4.12}$$

where the total energy E is a constant. This can be rearranged into the form:

$$-\frac{\hbar^2}{2\mu}\frac{d^2\psi}{dx^2} + \frac{1}{2}kx^2\psi = E\psi \tag{4.13}$$

As a trial solution of this differential equation we shall look at the bell-shaped wavefunction:

$$\psi_0 = N_0 e^{-ax^2} \tag{4.14}$$

Here, N_0 is a normalization constant, and a is another constant, the value of which we have to determine. Now:

$$\frac{d\psi_0}{dx} = -2axN_0 e^{-ax^2} \tag{4.15}$$

and

$$\frac{d^2\psi_0}{dx^2} = N_0(-2a + 4a^2 x^2)e^{-ax^2} \tag{4.16}$$

When this expression for $d^2\psi/dx^2$ is substituted into equation (4.13), we obtain:

$$\left(-\frac{\hbar^2}{2\mu}\right)N_0(-2a + 4a^2 x^2)e^{-ax^2} + \frac{1}{2}kx^2 N_0 e^{-ax^2} = E_0 N_0 e^{-ax^2} \tag{4.17}$$

where E_0 is the energy for the wavefunction ψ_0. The term $N_0 e^{-ax^2}$ is a common factor throughout this equation and cancels out. This leaves:

$$\left(-\frac{\hbar^2}{2\mu}\right)(-2a + 4a^2 x^2) + \frac{1}{2}kx^2 = E_0 \tag{4.18}$$

The total energy, E_0, must be a constant and cannot therefore vary with x. This will be true only if the two terms in x^2 cancel, which requires that:

$$\frac{k}{2} = \frac{2a^2\hbar^2}{\mu} \tag{4.19}$$

Thus, the wavefunction $\psi_0 = N_0 e^{-ax^2}$ will be a solution of the Schrödinger equation if:

$$a = \sqrt{\frac{k\mu}{4\hbar^2}} = \frac{1}{2\hbar}\sqrt{k\mu} \tag{4.20}$$

The energy is then given by the equation:

$$E_0 = \frac{a\hbar^2}{\mu} = \frac{\hbar}{2\mu}\sqrt{k\mu} = \frac{\hbar}{2}\sqrt{\frac{k}{\mu}} \tag{4.21}$$

When this equation is combined with the classical expression for the frequency (equation 4.11), we obtain the result:

$$E_0 = \tfrac{1}{2}h\omega_0 \tag{4.22}$$

This is known as the zero-point energy of the simple harmonic oscillator.

Other solutions of the wave equation can be found, all involving the factor e^{-ax^2} multiplied by a polynomial in x, known as a Hermite polynomial, and given the symbol H_v. There are an infinite number of these polynomials, and they are indexed by an integer, v, starting with $v = 0$ for the ground state. The first few wavefunctions and their associated energies are given in Table 4.1. For simplicity the normalization factors N_v have not been explicitly given.

Table 4.1　Wavefunctions and energies for the simple harmonic oscillator

Wavefunctions*	Energies
$\psi_0 = N_0 e^{-ax^2}$	$E_0 = \frac{1}{2} h\omega_0$
$\psi_1 = N_1 x e^{-ax^2}$	$E_1 = \frac{3}{2} h\omega_0$
$\psi_2 = N_2 (4ax^2 - 1) e^{-ax^2}$	$E_2 = \frac{5}{2} h\omega_0$
$\psi_v = N_v H_v(x) e^{-ax^2}$	$E_v = (v + \frac{1}{2}) h\omega_0$

$$^*a = \frac{\sqrt{k\mu}}{2\hbar} = \frac{\pi\sqrt{k\mu}}{h}$$

It can be seen that the energy levels are equally spaced and follow the formula:

$$E_v = \left(v + \frac{1}{2}\right) h\omega_0 \tag{4.23}$$

The wavefunctions and associated energies are shown in Figure 4.5, together with the parabolic potential energy curve. Each wavefunction has been given a baseline which corresponds to the total energy of the particle.

4.2.3 Application of Harmonic Oscillator Model to Infrared Spectra of Diatomic Molecules

For most chemical bonds the spacing between the vibrational energy levels is such that vibrational excitation of the molecules causes absorption of radiation in the infrared region of the spectrum. The potential energy curve for the vibration of a diatomic molecule is shown in Figure 4.6, together with the vibrational energy levels. The parabolic curve obtained from the harmonic oscillator model, and the associated energy levels, are shown as dotted lines. It can be seen that the fit to the actual data is quite good for $v = 0$ and $v = 1$, but becomes progressively worse for higher values of v.

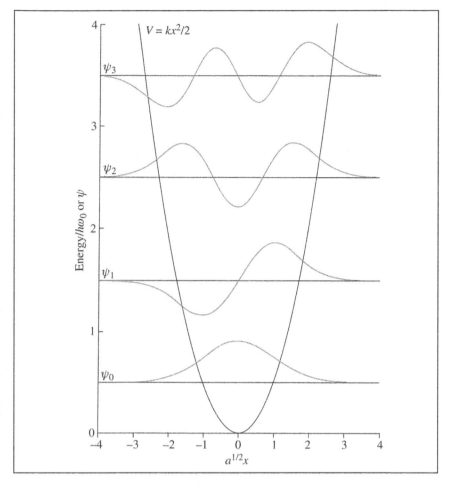

Figure 4.5 The wavefunctions, potential energy and total energies for a harmonic oscillator

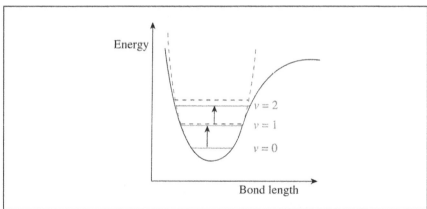

Figure 4.6 Comparison of the potential energy curve and total energy of a real diatomic oscillator (full lines) with those for the harmonic oscillator (dotted lines)

An analysis of the vibrational spectrum allows the force constant of the bond to be evaluated.

In spectroscopy it is normal to quote the frequency at which an absorption occurs in terms of a wavenumber. This is equal to the reciprocal wavelength and is given in units of cm^{-1}.

Worked Problem 4.2

Q The vibrational transition from $v = 0$ to $v = 1$ for carbon monoxide occurs at a wavenumber of 2143.3 cm^{-1}. Calculate a value for the force constant of the CO bond.

A We assume that the molecule behaves like a simple harmonic oscillator and that equations (4.11) and (4.23) can be used to calculate the energy levels. The energy required to excite the molecule from $v = 0$ to $v = 1$ is therefore:

$$\Delta E = h\omega_0 = \frac{h}{2\pi}\sqrt{\frac{k}{\mu}}$$

This must be equal to the energy of the photon, which is obtained from the equation:

$$E = h\nu = \frac{hc}{\lambda}$$

From these two equations we obtain the following expression for the force constant:

$$k = \frac{4\pi^2 c^2 \mu}{\lambda^2}$$

The reduced mass of CO is given by the expression

$$\mu = \frac{m_C m_O}{m_C + m_O}$$

The relative atomic masses of ^{12}C and ^{16}O are 12.01 and 16.00, respectively, and the atomic mass unit is 1.661×10^{-27} kg. Hence:

$$\mu = \frac{12.01 \times 16.00 \times 1.661 \times 10^{-27} \text{ kg}}{12.01 + 16.00} = 1.139 \times 10^{-26} \text{ kg}$$

The infrared absorption occurs at 2143.2 cm^{-1}, which is a reciprocal wavelength. Hence, $\lambda = 4.666 \times 10^{-6}$ m, and

$$k = \frac{4 \times (3.142)^2 \times (2.998 \times 10^8 \text{ m s}^{-1})^2 \times (1.139 \times 10^{-26} \text{ kg})}{(4.666 \times 10^{-6} \text{ m})^2}$$

$$= 1856 \text{ N m}^{-1}$$

This value is quite close to the accurately determined force constant of 1902 N m^{-1} (see Banwell and McCash[1]). The reason why this simple calculation does not give the true value is that the vibration of the CO molecule deviates slightly from simple harmonic motion, and an anharmonic factor has to be included in an accurate calculation to allow for this.

4.2.4 Extension of the Wavefunctions beyond Classical Limits

The wavefunctions shown in Figure 4.5 look very similar to the particle-in-a-box wavefunctions except that they extend beyond the classical limits of the potential well. In classical mechanics the maximum extension of the vibrating bond, x_{max}, occurs at the point where the total energy is equal to the potential energy, that is, when $\frac{1}{2}kx_{max}^2 = E$. This gives:

$$x_{max} = \pm \left(\frac{2E}{k} \right)^{1/2} \tag{4.24}$$

Thus, in Figure 4.5 the maximum classical extension of the bond occurs at the point where the horizontal, total energy line crosses the parabolic potential energy curve.

In quantum mechanics, however, there is a finite probability of finding the bond extended beyond this classical limit. In such a region the kinetic energy would appear to be negative because the total energy E is now less than the potential energy V. An examination of the one-dimensional Schrödinger equation:

$$-\frac{\hbar^2}{2m} \left(\frac{d^2\psi}{dx^2} \right) = (E - V)\psi \tag{4.25}$$

shows that $d^2\psi/dx^2$ has the same sign as ψ when $E - V$ is negative. Under these conditions the wavefunction no longer oscillates, but exhibits an exponential decay. This can be seen quite clearly for the three wavefunctions shown in Figure 4.5. Between the classical limits, the wavefunctions oscillate because $d^2\psi/dx^2$ has the opposite sign from ψ, but at distances greater than x_{max} they decay exponentially to zero.

4.3   Quantum Mechanical Tunnelling

The existence of wavefunctions in regions where $E < V$ leads to a process known as tunnelling. This enables light particles, such as electrons and protons, to penetrate regions of space that should be forbidden to them according to the laws of classical mechanics.

4.3.1 Equation for Tunnelling through a One-dimensional Rectangular Barrier

We will start with the simple model shown in Figure 4.7. Here, a particle with total energy E approaches a rectangular barrier, where the potential energy rises suddenly from zero to a value V_0, which is greater than E. Classically, the particle would be unable to penetrate the barrier because it does not have sufficient energy, and it would be totally reflected at the boundary. However, in quantum mechanics there is a finite probability that the particle will appear on the other side of the barrier by the process of tunnelling.

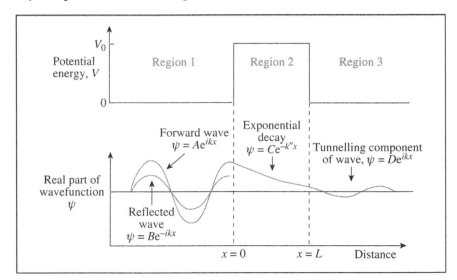

Figure 4.7 Wavefunctions associated with the tunnelling of a particle through a rectangular potential energy barrier

The one-dimensional Schrödinger equation to be solved is:

$$-\frac{\hbar^2}{2m}\frac{\mathrm{d}^2\psi}{\mathrm{d}x^2} = (E - V)\psi \qquad (4.26)$$

There are now two types of solution, depending on whether $(E - V)$ is +ve or −ve.

Region 1: in Front of the Barrier

We will denote the wavefunction in this region as ψ_1. Since $V = 0$, the wave equation becomes:

$$-\frac{\hbar^2}{2m}\frac{\mathrm{d}^2\psi_1}{\mathrm{d}x^2} = E\psi_1 \tag{4.27}$$

The general solution of this equation is:

$$\psi_1 = A\mathrm{e}^{ikx} + B\mathrm{e}^{-ikx} \tag{4.28}$$

where A, B and k are constants. The first part of the wavefunction is the incident wave, representing the particle approaching the barrier, and the second part is the reflected wave. Since the probability of the particle tunnelling through the barrier is normally very small, the reflected wave is going to be almost as strong as the incident wave. By differentiating ψ_1 twice with respect to x, and comparing the result with equation (4.27), it can be shown that:

$$k = \frac{\sqrt{2mE}}{\hbar} \tag{4.29}$$

Region 2: Inside the Barrier

In this region, $(E - V_0)$ is negative, and the wave equation can be written as:

$$\frac{\hbar^2}{2m}\frac{\mathrm{d}^2\psi_2}{\mathrm{d}x^2} = \left(V_0 - E\right)\psi_2 \tag{4.30}$$

A solution of this equation, which is appropriate for thick barriers, is:

$$\psi_2 = C\mathrm{e}^{-k''x} \tag{4.31}$$

where C and k'' are constants. It should be noted that "i" is not needed in the exponent because V_0 is now greater than E. The absence of "i" means that ψ_2 does not represent an oscillating wave, but simply an exponential decay function, as shown in Figure 4.7. By differentiating ψ_2 twice with respect to x, and comparing the result with equation (4.30), it can be shown that:

$$k'' = \frac{\sqrt{2m(V_0 - E)}}{\hbar} \tag{4.32}$$

Region 3: Beyond the Barrier

In this region, $V = 0$, and the wave equation has the same form as equation (4.27). One difference is that we have to consider only the particle moving away from the barrier, because there are no particles approaching the barrier from the other side. The wavefunction can therefore be written as:

$$\psi_3 = De^{ikx} \tag{4.33}$$

where D is another constant.

The total wavefunction is obtained by joining together the separate wavefunctions from the three regions. Ideally, this should be done in such a way that there is no discontinuity in either ψ or $d\psi/dx$ at the boundaries between the regions. Thus, at the boundary between regions 1 and 2, we have $\psi_1 = \psi_2$ and $d\psi_1/dx = d\psi_2/dx$, and similar conditions apply at the boundary between regions 2 and 3. An example of how the real parts of the wavefunctions can be joined up is shown in Figure 4.8.

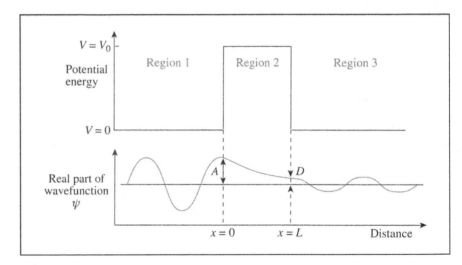

Figure 4.8 A simplified model for the tunnelling of a particle through a rectangular potential energy barrier

To simplify the mathematics we will make two approximations. The first is that the reflected component of ψ_1, Be^{-ikx}, can be ignored. The second is that we will only attempt to match the magnitudes, and not the gradients, of the wavefunctions at the boundaries. At $x = 0$, $\psi_1 = A$ and $\psi_2 = C$, and the boundary condition requires that:

$$A = C \tag{4.34}$$

At $x = L$ the boundary condition requires that:

$$Ce^{-k''L} = De^{ikL} \tag{4.35}$$

Combination of equations (4.34) and (4.35) gives:

$$D = Ae^{-L\left(k''+ik\right)} \tag{4.36}$$

The rate at which particles leave the barrier after tunnelling is proportional to $(D^*e^{-ikx})(De^{ikx})$ and the rate at which particles arrive at the front of the barrier is proportional to $(A^*e^{-ikx})(Ae^{ikx})$. The asterisk denotes the complex conjugate and is a reminder that the constants A and D can be complex. The tunnelling probability, P, is equal to the ratio of these two terms. Thus:

$$P = \frac{D^* De^{-ikx}e^{ikx}}{A^* Ae^{-ikx}e^{ikx}} = \frac{D^* D}{A^* A} \tag{4.37}$$

Equation (4.36) is now combined with this equation to obtain the following expression for the tunnelling probability:

$$P = e^{-L\left(k''+ik\right)}e^{-L\left(k''-ik\right)} = e^{-2Lk''} \tag{4.38}$$

By substitution of the expression for k'' given in equation (4.32), this equation becomes:

$$P = \exp\left\{\frac{-2L[2m\left(V_0 - E\right)]^{1/2}}{\hbar}\right\} \tag{4.39}$$

Because of the approximations that have been made, this equation is useful only for obtaining the order of magnitude of the tunnelling probability, although it does contain the dominant term. More accurate expressions differ by having a factor involving E/V_0 in front of the exponential term.

The tunnelling probability is found to be exponentially dependent upon the length and height of the potential energy barrier, and also on the mass of the particle. How this comes about can be understood by referring to Figure 4.8. Inside the barrier the wavefunction decays away exponentially, and it is the value of ψ at the end of the barrier that determines the strength of the oscillatory wave that emerges. Thus, as the barrier becomes longer, the tunnelling probability drops off very rapidly. The height of the barrier, $(V_0 - E)$, and the mass of the particle also affect the rate at which the wavefunction decays away with distance. For a barrier of fixed dimensions, the kinetic energy of the approaching particle and its mass will be very important in determining the overall tunnelling probability. In general, the wavelength of the approaching particle has to be of the same order of magnitude as the length of the barrier for tunnelling to be important. Tunnelling is therefore likely to occur only for relatively light particles such as electrons and, to a lesser extent, protons. At low temperature, protons are known to diffuse through metals such as tantalum by a process of tunnelling[2] and, as we shall see in Section 4.3.4, electron tunnelling is the basis of the important technique of scanning tunnelling microscopy.

4.3.2 Can Particles really have Negative Kinetic Energy?

There is a finite probability of finding the particle inside the barrier, where $(E - V)$ is negative, but this does not mean that the particle can exist there with negative kinetic energy. The mere act of locating the particle at a specific point in this classically forbidden region will drastically alter the wavefunction, and introduce a large uncertainty into the total energy of the particle, which is then going to be greater than V_0. In its original state, with total energy E less than V_0, the particle/wave entity is not localized at any particular point, and it is better to think of it predominantly in wave terms.

4.3.3 Tunnelling of Electrons

Figure 4.9 shows the potential energy barrier existing between two pieces of the same metal, separated by a distance of the order of one nanometre and placed in a vacuum. The highest occupied energy level in the metal is known as the Fermi level, and has an energy E_F. The energy of an electron just outside one of the pieces of metal is equal to V_0, the vacuum level, and this corresponds to the top of the barrier. The amount of energy required to take an electron from the Fermi level to the vacuum level is known as the work function of the metal and is given the symbol ϕ.

Figure 4.9a shows the situation where the two pieces of metal are at the same potential. Under these conditions, tunnelling cannot occur because there are no unoccupied energy levels on the other side of the barrier into which the tunnelling electrons can go. In Figure 4.9b, a small potential, V, has been applied so as to make the metal on the right more positive than the metal on the left. This allows electrons to tunnel from energy levels close to the Fermi level on the left side into vacant energy levels on the right side. For small, applied potentials, the potential energy barrier remains roughly rectangular, and the tunnelling probability

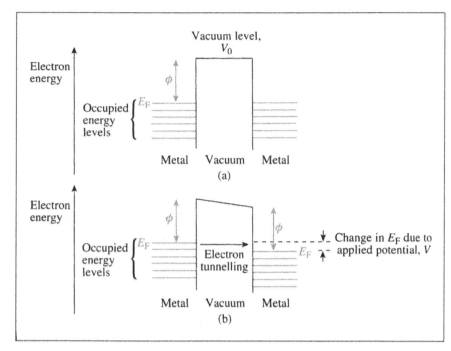

Figure 4.9 Diagram of the potential energy barrier existing between two pieces of metal: (a) both metals at the same potential; (b) with a small potential difference applied between the metals

for an electron approaching the gap can be estimated by using equation (4.39). For most metals the work function has a value between 4 and 5 eV, and when a value of 4.5 eV is substituted into equation (4.39), a tunnelling probability of about 3×10^{-10} is calculated for a one nanometre gap. The probability is extremely sensitive to the distance: if the gap between the two pieces of metal is increased to 1.1 nm, the tunnelling probability drops to about 4×10^{-11}.

Worked Problem 4.3

Q The electron tunnelling current flowing from a metal tip to a conducting surface, both held at fixed potentials, is 1.0×10^{-9} A when the separation between them is 0.6 nm. Using a rectangular potential energy barrier, estimate the current that will flow if the distance is reduced to 0.5 nm. The work function of the metal tip is equal to 4.5 eV.

A The work function of the metal, ϕ, can be obtained in units of Joules as follows:

$$\phi = \left(4.5\,eV\right)\left(1.6 \times 10^{-19}\,J\,[eV]^{-1}\right) = 7.2 \times 10^{-19}\,J$$

This is equal to $V_0 - E$. The tunnelling probability is then found from equation (4.39) to be:

$$P = \exp\left\{ \frac{-2\times\left(6\times10^{-10}\text{ m}\right)\times\left[2\times\left(9.1\times10^{-31}\text{ kg}\right)\times\left(7.2\times10^{-19}\text{ J}\right)\right]^{1/2}}{\left(1.05\times10^{-34}\text{ J s}\right)} \right\}$$

$$= \exp(-13.1) = 2.1\times10^{-6}$$

If the distance is reduced to 0.5 nm the probability becomes:

$$P = \exp(-10.9) = 1.8\times10^{-5}$$

Thus, the tunnelling probability has been increased by a factor of 8.6, and the new current will be approximately 8.6×10^{-9} A.

4.3.4 Scanning Tunnelling Microscopy (STM)

Binnig and Rohrer[3] received the Nobel Prize in 1986 for developing this important technique, which gives atomic resolution of surfaces. A schematic diagram of their set-up is shown in Figure 4.10. A very fine metal tip is lowered towards a conducting or semiconducting surface until electrons are able to tunnel between the tip and the surface. Typically, a bias voltage of 0.1–1.0 V is applied, and tunnelling occurs when the tip and surface are separated by distances of between 0.3 and 1.0 nm. Tunnelling currents are quite small, measuring between 0.1 and 1.0 nA.

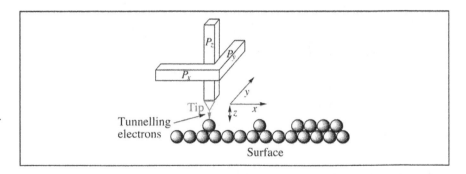

Figure 4.10 Method of operation of the scanning tunnelling microscope. P_x, P_y and P_z are piezoelectric ceramics

The tip consists of a cluster of atoms, and usually one atom sticks out more than the others; it is this atom which is responsible for most of the tunnelling current.

As shown in Figure 4.10, the tip is attached to a piezoelectric ceramic, a material that will expand in one direction and contract in another when a current is passed. By changing the current flowing through the ceramic in different directions, the tip can be either raised/lowered or slowly moved horizontally across the surface. In normal operation the tip is scanned across the surface, keeping the tunnelling current, and therefore the separation from the surface, constant (see Figure 4.11).

The tip is scanned across the surface in much the same way that an electron beam is used to form a television picture.

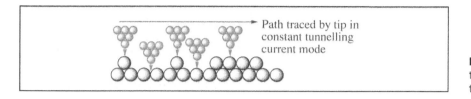

Path traced by tip in constant tunnelling current mode

Figure 4.11 Movement of the tip across the surface when the tunnelling current is kept constant

The varying piezoelectric currents are stored in a computer and are used to generate a topographical map of the surface. Because the current is very sensitive to the separation between tip and surface, this technique gives a very accurate profile of the surface. Atomic resolution can be attained, as shown by the STM image of gallium arsenide[4] seen in Figure 4.12.

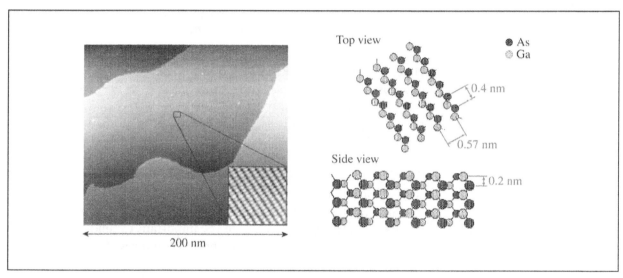

Top view

● As
◐ Ga

0.4 nm

0.57 nm

Side view

0.2 nm

200 nm

Figure 4.12 Comparison of the STM image of a gallium arsenide surface with a model surface. The main image shows only the steps that occur on the GaAs surface, but the high-resolution inset gives atomic resolution of a small part of the surface. The image was obtained with the tip positively biased with respect to the surface so that electrons tunnel from the gallium arsenide to the tip. In this mode the arsenic atoms are imaged. If the bias is reversed, so that electrons tunnel from tip to surface, the gallium atoms can be imaged

Summary of Key Points

1. The one-dimensional Schrödinger equation can be written as:

$$-\frac{\hbar^2}{2m}\left(\frac{d^2\psi}{dx^2}\right) = (E - V)\psi$$

Here, ψ is the wavefunction, m is the mass of a particle moving along the x axis, E is the total energy and V is the potential energy. This equation produces an oscillatory wave when E is greater than V, but generates an exponential decay function when E is less than V.

2. When this equation is applied to a one-dimensional harmonic oscillator, it produces a series of wavefunctions with energies given by the following formula:

$$E_v = \left(v + \frac{1}{2}\right)h\omega_0$$

Here, ω_0 is the fundamental frequency of the oscillator, and v is a quantum number that can have values of 0, 1, 2, *etc*. The oscillator has a minimum energy of $\frac{1}{2}h\omega_0$, and this is known as the zero-point energy. These concepts have been applied to the vibrational spectra of diatomic molecules.

3. Particles with small mass (*e.g.* protons and electrons) are able to pass through narrow energy barriers where the total energy is less than the potential energy. This process is known as tunnelling. Electron tunnelling is the basis of the important technique of scanning tunnelling microscopy, which produces images of surfaces with atomic resolution.

Problems

4.1. Show that the wavefunction $\psi = A\sin(kx) + B\cos(kx)$, where A, B and k are constants, is a solution of the one-dimensional Schrödinger equation, and hence derive an expression for the energy.

4.2. Calculate the zero-point vibrational energies of $^1H^{35}Cl$ and

$^2D^{35}Cl$ in units of kJ mol^{-1}, assuming that the molecules undergo simple harmonic oscillation and that the force constant of each bond is equal to 516 N m^{-1}.

4.3. In the infrared spectrum of nitric oxide the vibrational transition from $v = 0$ to $v = 1$ is observed at a wavelength of 5.330 $\times 10^{-6}$ m. Use the simple harmonic oscillator model to calculate the force constant of the bond and compare your answer with the accurately determined value of 1595 N m^{-1}.

4.4. An otherwise flat conducting surface has a patch of conducting material on it which is 0.3 nm thick. A negatively biased metal tip with a work function of 4.0 eV is used to scan this system at a constant height of 0.6 nm above the general level of the surface. Calculate the factor by which the tunnelling current increases as the tip passes over the patch.

References

1. C. N. Banwell and E. M. McCash, *Fundamentals of Molecular Spectroscopy*, 4th edn., McGraw-Hill, London, 1994.
2. J. M. Guil, D. O. Hayward and N. Taylor, *Proc. R. Soc. London, Ser. A*, 1973, **335**, 141.
3. G. Binnig and H. Rohrer, *Surf. Sci.*, 1983, **126**, 236.
4. D. M. Holmes, J. G. Belk, J. L. Sudijono, J. H. Neave, T. S. Jones and B. A. Joyce, *Surf. Sci*, 1995, **341**, 133.

Further Reading

R. P. Bell, *The Proton in Chemistry*, 2nd edn., Chapman & Hall, London, 1973.

R. Wiesendanger, *Scanning Probe Microscopy and Spectroscopy: Methods and Applications*, Cambridge University Press, Cambridge, 1994.

5

Rotational Motion

Aims

By the end of this chapter you should be able to:

*   Use the one-dimensional Schrödinger equation to derive the wavefunctions for a particle moving in a circle
*   Apply the three-dimensional Schrödinger equation to the movement of a particle on the surface of a sphere
*   Apply the spherical model to the rotation of diatomic molecules and to electron and nuclear spin

All the systems that we have considered so far have been restricted to motion along a straight line, whereas most of the systems of interest to chemists involve particles circulating around a fixed point, for example electrons circulating around nuclei, molecules rotating about their centre of mass and electrons spinning on their axes. In this chapter we consider those systems in which a particle rotates at a constant distance from a fixed point. In the following chapter this treatment will be extended to the hydrogen atom, where the distance of the electron from the nucleus is variable.

5.1 Circular Motion in a Fixed Plane

Vector notation. In this chapter we meet vectors for the first time. The velocity, v, is a vector quantity because it has both magnitude and direction. Such vectors will normally be shown in bold face type, and ordinary face type will be reserved for situations where only the magnitude of the vector is required. Thus, v will denote the speed of a particle. The square of a vector is automatically defined as a scalar quantity, being equal to the magnitude of the vector squared, and it will always be shown in ordinary type.

We begin with the simplest case in which a particle of mass m moves at a constant speed v around a circle of radius r, as shown in Figure 5.1. The distance moved by the particle along the circumference of the circle from its starting position is denoted by s, and the angle subtended at the centre is denoted by ϕ. It should be noted that s and ϕ are considered to be positive for anti-clockwise motion and negative for clockwise motion. Before obtaining solutions of the Schrödinger equation for this system, it will be useful to review some basic concepts from classical mechanics.

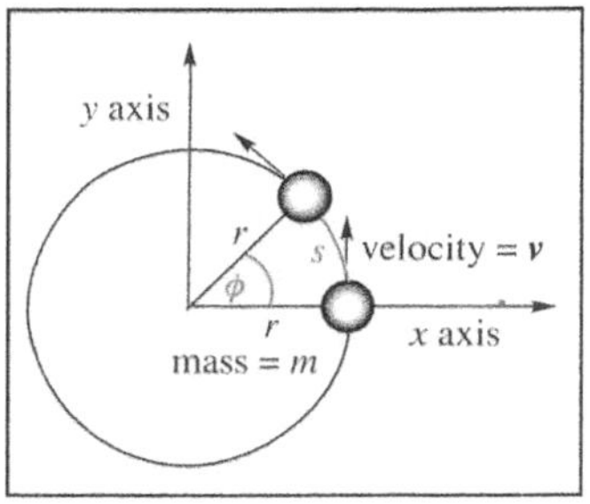

5.1.1 Classical Treatment

The velocity of the particle at any instant is represented by the vector v. Whilst the magnitude of this vector remains constant, its direction is continually changing as the particle moves around the circle, and it is not a constant of the motion. This difficulty can be overcome by describing the movement in terms of the angular velocity ω, which is numerically equal to the rate of change of the angle ϕ. The angular velocity is a vector which points in a direction at right angles to the plane of motion, and therefore does not change direction as the particle goes around the circle. To distinguish between clockwise and anti-clockwise motion, the vector points up or down according to the right-hand screw rule, which is illustrated in Figure 5.2.

Figure 5.1 Movement of a particle around a circle

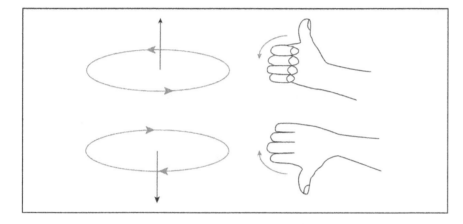

Figure 5.2 The right-hand screw rule for determining the direction of the angular velocity or angular momentum vector

Because $\phi = s/r$ and $ds/dt = v$, it is possible to write the following expression for the magnitude of the angular velocity:

$$\omega = \frac{d\phi}{dt} = \frac{1}{r}\frac{ds}{dt} = \frac{v}{r} \tag{5.1}$$

The kinetic energy of the particle is constant, and equal to $mv^2/2$. In terms of the angular velocity ω, this can be expressed as:

$$E = \frac{1}{2}mr^2\omega^2 \tag{5.2}$$

The quantity mr^2 is known as the moment of inertia of the particle and is given the symbol I. Thus:

$$E = \frac{1}{2}I\omega^2 \tag{5.3}$$

This equation has the same form as the equation for linear motion, but with moment of inertia and angular velocity substituted for mass and linear velocity, respectively. For rotational motion it is generally true that the angular velocity and the moment of inertia have an analogous role to velocity and mass in linear motion. In keeping with this principle, we can define an angular momentum vector, L, as follows:

$$L = I\omega \tag{5.4}$$

Clearly, this vector has the same directional properties as ω. In terms of L, equation (5.3) then becomes:

$$E = \frac{L^2}{2I} \tag{5.5}$$

This equation is equivalent to the equation $E = p^2/2m$ for linear motion. With the substitutions $I = mr^2$ and $\omega = v/r$, the magnitude of the angular momentum can also be written as:

$$L = mvr \tag{5.6}$$

5.1.2 Quantum Mechanical Treatment

The one-dimensional Schrödinger equation can still be used to describe the motion of the particle if the Cartesian coordinate x is replaced by s, the distance moved by the particle along the circumference of the circle from its starting point. Although this variable describes a curved path, its use in place of x can be justified because the particle is constrained to move along this path. With the potential energy V put equal to zero, the Schrödinger equation can then be written as:

$$-\frac{\hbar^2}{2m}\left(\frac{d^2\psi}{ds^2}\right) = E\psi \tag{5.7}$$

The length of the arc of a circle is equal to the radius of the circle multiplied by the angle subtended at the centre, that is, $s = r\phi$. Thus, $ds = rd\phi$, and equation (5.7) becomes:

$$-\frac{\hbar^2}{2mr^2}\left(\frac{d^2\psi}{d\phi^2}\right) = E\psi \tag{5.8}$$

This equation could have been obtained more rigorously by solving the two-dimensional Schrödinger equation:

$$-\frac{\hbar^2}{2m}\left(\frac{\partial^2\psi}{\partial x^2}+\frac{\partial^2\psi}{\partial y^2}\right)=E\psi \qquad (5.9)$$

and making the substitutions $x = r\cos\phi$ and $y = r\sin\phi$, but this method involves much more complicated mathematics.

The moment of inertia, I, of the rotating particle is defined by the relation $I = mr^2$. Thus, equation (5.8) can be rewritten as:

$$-\frac{\hbar^2}{2I}\left(\frac{d^2\psi}{d\phi^2}\right)=E\psi \qquad (5.10)$$

From equation (5.5) the energy is equal to $L^2/2I$, where L is the angular momentum of the particle; equation (5.10) can therefore be rearranged to give:

It is worth noting that equation (5.10) is identical to the one-dimensional Schrödinger equation for linear motion, except that the Cartesian coordinate x has been replaced by the angle ϕ and the mass has been replaced by the moment of inertia I.

$$-\frac{d^2\psi}{d\phi^2}=\left(\frac{L}{\hbar}\right)^2\psi \qquad (5.11)$$

Putting $L/\hbar = \alpha$, the solutions of this equation take the form:

$$\psi = Ne^{\pm i\alpha\phi} \qquad (5.12)$$

where N is a normalization constant.

If the wavefunction is expanded in the form:

$$\psi = N(\cos\alpha\phi \pm i\sin\alpha\phi) \qquad (5.13)$$

it can be seen to consist of two sinusoidal waves, one real and the other complex.

Boundary Condition

The boundary condition relevant to motion in a circle is different from that required for a particle in a box, where the wavefunction had to go to zero at the ends of the box. For circular motion the wavefunction has to match up with itself after one complete revolution of the circle. This requires the circumference of the circle to be equal to a whole number of wavelengths. The situation where five wavelengths fit into the circle is illustrated for the sine function in Figure 5.3a. The plot for the cosine function would be similar, but rotated through 90°. If this condition is not met the waves will not coincide with one another after one complete revolution, and multiple values of ψ will be obtained for any particular point on the circle, as shown in Figure 5.3b. As we saw in Section 1.4.5,

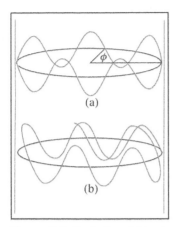

Figure 5.3 Illustration of (a) satisfactory and (b) unsatisfactory wavefunctions. In the first case, five wavelengths fit exactly along the perimeter of the circle, but in the second case there is a mismatch

satisfactory wavefunctions have to be single-valued, so only wavefunctions of the type illustrated in Figure 5.3a are allowed.

This can be expressed in mathematical terms by stating that the magnitude of the wavefunction must remain the same when the angle ϕ is increased by an amount 2π, that is:

$$\sin\alpha\phi = \sin\alpha(\phi + 2\pi) = \sin(\alpha\phi + \alpha 2\pi) \qquad (5.14)$$

and:

$$\cos\alpha\phi = \cos\alpha(\phi + 2\pi) = \cos(\alpha\phi + \alpha 2\pi) \qquad (5.15)$$

This will be true only if α is equal to zero or an integer (quantum number), which we will call m_l. Thus, acceptable wavefunctions take the form:

$$\psi = Ne^{im_l\phi} \quad m_l = 0, \pm 1, \pm 2, \pm 3, \dots \qquad (5.16)$$

Positive values of m_l represent anti-clockwise motion, whereas negative values represent clockwise motion. When $m_l = 0$ the particle is stationary. The real (cosine) parts of the wavefunctions for $m_l = 0, \pm 1$ are shown in Figure 5.4. The complex (sine) parts will be similar but rotated through $90°$.

Figure 5.4 The real (cosine) parts of the wavefunction for a particle moving in a circle

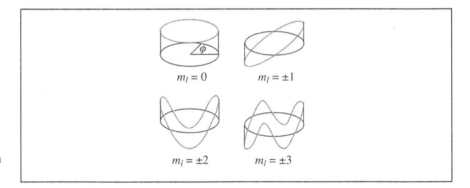

$m_l = 0$ $m_l = \pm 1$

$m_l = \pm 2$ $m_l = \pm 3$

Worked Problem 5.1

Q Evaluate the normalization constant N in equation (5.16).

A The probability of finding the particle between ϕ and $\phi + d\phi$ on the circumference of the circle is equal to $\psi^*\psi d\phi$. Since the particle must be somewhere on the circle, the integral of this expression over an angle of 2π must be equal to one. Hence:

$$N^2 \int_{\phi=0}^{\phi=2\pi} e^{-im_l\phi} e^{im_l\phi} d\phi = N^2 \int_{\phi=0}^{\phi=2\pi} d\phi = 1$$

This equation is satisfied when $N = \dfrac{1}{\sqrt{2\pi}}$. Thus:

$$\psi = \frac{1}{\sqrt{2\pi}} e^{im_l\phi} \quad m_l = 0, \pm 1, \pm 2, \pm 3, \dots \qquad (5.17)$$

5.1.3 Quantizaton of Angular Momentum and Energy

Remembering that $m_l = \alpha = L/\hbar$, we see that the magnitude of the angular momentum of the particle is quantized in units of $\hbar$, that is:

$$L = m_l\hbar \quad m_l = 0, \pm 1, \pm 2, \pm 3, \dots \qquad (5.18)$$

Thus, we come to the important conclusion that the angular momentum of a particle moving at constant speed in a circle can only have values which are multiples of $\hbar$. This quantization in units of $\hbar$ was first postulated by Niels Bohr in 1913 for the movement of an electron around the hydrogen atom. Later, we shall see that it applies quite generally to the movement of electrons in atoms and molecules.

From equation (5.5) the energy can be expressed as:

$$E = \frac{\hbar^2 m_l^2}{2I} \quad m_l = 0, \pm 1, \pm 2, \pm 3, . \qquad (5.19)$$

The energy levels are shown in Figure 5.5. It can seen that, except for the ground state, there are always two states with the same energy, corresponding to clockwise and anti-clockwise movement. Such states are said to be doubly degenerate.

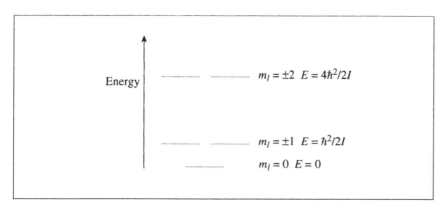

Figure 5.5 The energy levels for a particle moving in a circle

Q Use the de Broglie relation to show that the angular momentum of a particle moving in a circle is quantized in units of $\hbar$.

A Since the circumference of the circular path has to be equal to an integral number of wavelengths, we can write:

$$2\pi r = m_l \lambda$$

where the wavelength is taken to be negative when the particle is moving in a clockwise direction. When this equation is combined with the de Broglie relation, $p = h/\lambda$, the following expression is obtained for the magnitude of the linear momentum at any instant:

$$p = \frac{m_l h}{2\pi r}$$

It can be seen from equation (5.6) that the magnitude of the angular momentum L is equal to mvr, and so to pr. Hence:

$$L = \frac{m_l h}{2\pi} = m_l \hbar$$

5.2   Rotation in Three Dimensions

In this section we consider a particle that is free to move in three dimensions, but always at a constant distance r from a fixed point. In this way, motion is confined to the surface of a sphere, as illustrated in Figure 5.6. The potential energy V is also constant and can be set equal to zero. This model can be applied to the rotation of rigid molecules and also to electron and nuclear spin, as we shall see later. However, the main reason for introducing it at this stage is that it provides a first step towards obtaining mathematical solutions of the Schrödinger equation for the hydrogen atom.

5.2.1 The Schrödinger Equation in Spherical Polar Coordinates

The Schrödinger equation given in Section 4.1 can be extended to three dimensions by writing:

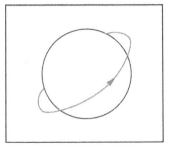

Figure 5.6 Movement of a particle on the surface of sphere

$$-\frac{\hbar^2}{2m}\nabla^2\psi + V\psi = E\psi \qquad (5.20)$$

where:

$$\nabla^2 = \frac{\partial^2}{\partial x^2} + \frac{\partial^2}{\partial y^2} + \frac{\partial^2}{\partial z^2} \qquad (5.21)$$

∇^2 is known as the Laplacian operator.

This form of the equation is not easily applied to rotational motion because the Cartesian coordinates used do not reflect the centro-symmetric nature of the problem. It is better to express the Schrödinger equation in terms of the spherical polar coordinates r, θ and ϕ, which are shown in Figure 5.7. Their mathematical relationship to x, y and z is given on the left of the diagram. In terms of these coordinates the Laplacian operator ∇^2 becomes:

$$\nabla^2 = \frac{1}{r^2}\left[\frac{\partial}{\partial r}\left(r^2\frac{\partial}{\partial r}\right) + \Lambda^2\right] \qquad (5.22)$$

where:

$$\Lambda^2 = \frac{1}{\sin\theta}\frac{\partial}{\partial\theta}\left(\sin\theta\frac{\partial}{\partial\theta}\right) + \frac{1}{\sin^2\theta}\frac{\partial^2}{\partial\phi^2} \qquad (5.23)$$

Although the Schrödinger equation looks much more formidable in spherical polar coordinates than it did in Cartesian coordinates, it is easier to solve because the wavefunctions can often be written as the product of three functions, each one of which involves only one of the variables r, θ and ϕ.

The symbol ∂ indicates differentiation with respect to one variable, while keeping the other two variables constant. This process is known as partial differentiation.

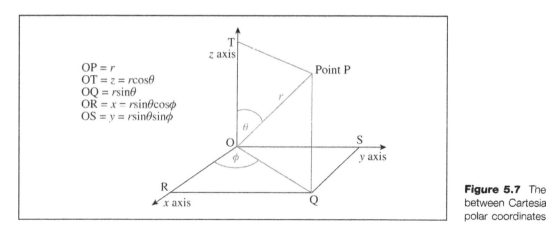

OP = r
OT = z = $r\cos\theta$
OQ = $r\sin\theta$
OR = x = $r\sin\theta\cos\phi$
OS = y = $r\sin\theta\sin\phi$

Figure 5.7 The relationship between Cartesian and spherical polar coordinates

Worked Problem 5.3

Q For circular motion in a fixed plane, show that the Laplacian operator given in equations (5.22) and (5.23) leads to the Schrödinger equation derived in Section 5.1.2.

A For a particle moving around a circle in the xy plane, the angle θ is fixed at 90° and r is also a constant. Thus, the differentials with respect to r and θ in equations (5.22) and (5.23) disappear. With $\sin\theta = 1$, the equation for ∇^2 simplifies to:

$$\nabla^2 = \frac{1}{r^2}\frac{d^2}{d\phi^2}$$

With $V = 0$, the Schrödinger equation then becomes:

$$-\frac{\hbar^2}{2mr^2}\frac{d^2\psi}{d\phi^2} = E\psi$$

which is the same as equation (5.8).

5.2.2 Solutions of the Schrödinger Equation for Rotation with Constant r

Because r is a constant, differentiation with respect to r can be ignored in equation (5.22) and the Schrödinger equation in spherical polar coordinates simplifies to:

$$-\frac{\hbar^2}{2mr^2}\Lambda^2\psi = E\psi \qquad (5.24)$$

It should be noted that mr^2 is equal to the moment of inertia of the particle, I.

Solutions of this equation take the form:

$$\psi = \Theta(\theta)\Phi(\phi) \qquad (5.25)$$

where Θ is a function only of the variable θ, and Φ is a function only of the variable ϕ. The mathematics involved in obtaining the Θ and Φ functions are rather lengthy and only the results will be given here. Those interested in the mathematics should refer to the Appendix at the end of the chapter.

The function Φ is found to be identical to the one already discussed for rotational motion in a fixed plane, that is:

$$\Phi = \frac{1}{\sqrt{2\pi}} e^{im_l \phi} \quad m_l = 0, \pm 1, \pm 2, \pm 3 \qquad (5.26)$$

The function Θ can take on various forms, depending on the values of two quantum numbers. The first is m_l, and the second is a new quantum number, l, sometimes known as the azimuthal quantum number. The latter is restricted to the values $l = 0, 1, 2, 3, \dots$

The modulus of m_l (that is, its magnitude regardless of sign) must never be greater than l. This can be written mathematically as:

$$|m_l| \leq l \qquad (5.27)$$

Thus, the allowed values of the two quantum numbers are as follows:

$l = 0$ $m_l = 0$
$l = 1$ $m_l = -1, 0, +1$
$l = 2$ $m_l = -2, -1, 0, +1, +2, \text{ etc.}$

The wavefunctions are called spherical harmonics because they represent the types of waveform that can be sustained on the surface of a sphere (imagine a tidal wave on a flooded planet), and they are given the symbol $Y_{l,m_l}(\theta, \phi)$ where the subscripts identify the quantum numbers controlling the mathematical form taken by the wavefunctions.

The first few spherical harmonics are shown in Figure 5.8, where only the real parts of the wavefunctions (represented by $\Theta \cos m_l \phi$) have been drawn. With respect to the real parts, the complex parts (represented by $\Theta \sin m_l \phi$) would be rotated by 90° in the xy plane. The lines separating +ve and –ve regions are nodes, points where the real component of the wavefunction is zero. The direction of motion will be at right angles to the nodal lines because this will be the direction in which the wavefunction has maximum curvature. So, for example, the motion for $l = 1$ and $m_l = 0$ will be entirely from one "pole" to the other, with no motion around the "equator". For $l = 1$ and $m_l = \pm 1$ the motion will be a mixture of movement between the "poles" and around the "equator". The more nodes that cut the "equator", the higher is the angular momentum of the particle in the z direction.

The real parts of the complex wavefunctions can also be regarded as wavefunctions in their own right, because they can be generated by a linear combination of complex wavefunctions with the same energy. Thus:

$$\sin\theta \, (e^{+i\phi} + e^{-i\phi}) = 2 \sin\theta \cos\phi \qquad (5.28)$$

and:

Those with some knowledge of chemical bonding will recognize that the spherical harmonics are related to the atomic orbitals, with $l = 0$, 1 and 2 corresponding to s, p and d orbitals, respectively. This subject will be developed further in Chapter 6.

Quantum numbers l	m_l	$\Theta\Phi$ (normalization constant omitted)	Sign of real part of wavefunction on surface of sphere
0	0	constant	
1	0	$\cos\theta$	
1	+1	$-\sin\theta\,e^{i\phi}$	
1	−1	$\sin\theta\,e^{-i\phi}$	
2	0	$3\cos^2\theta-1$	
2	+1	$-\cos\theta\sin\theta\,e^{i\phi}$	
2	−1	$\cos\theta\sin\theta\,e^{-i\phi}$	
2	+2	$\sin^2\theta\,e^{2i\phi}$	
2	−2	$\sin^2\theta\,e^{-2i\phi}$	

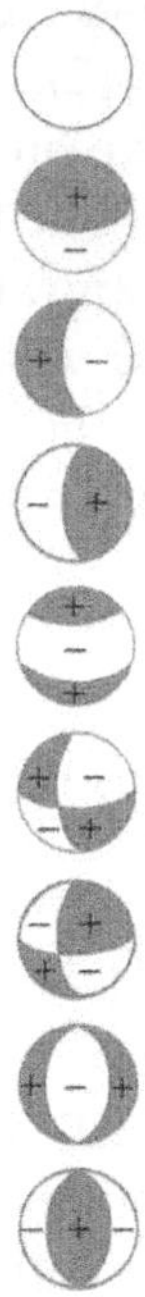

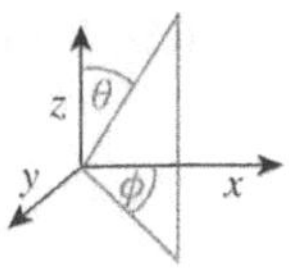

Figure 5.8 The first nine spherical harmonic wavefunctions

It should be noted that wavefunctions can be combined in this way only when they have the same energy. Such wavefunctions are said to be degenerate. It is generally true that any linear combination of degenerate wavefunctions will be a solution of the corresponding Schrödinger equation.

$$\sin\theta\,(e^{+i\phi} - e^{-i\phi}) = 2i\,\sin\theta\,\sin\phi \qquad (5.29)$$

Polar plots of these real wavefunctions are shown in Figure 5.9 for $l = 0$ and $l = 1$. In these diagrams the distance from the origin is proportional to the magnitude of the wavefunction. It can be seen that the three wavefunctions corresponding to $l = 1$ all have the same dumbbell shape, with positive and negative lobes, but oriented along different Cartesian axes. Once again, those with some knowledge of chemical bonding will recognize these shapes as those of the $2p_x$, $2p_y$ and $2p_z$ atomic orbitals.

5.2.3 Energies of the Wavefunctions

It is shown in the Appendix that the spherical harmonic wavefunctions, Y_{l,m_l}, have the following important property when operated upon by the operator Λ^2:

$$\Lambda^2 Y_{l,m_l} = -l\left(l+1\right)Y_{l,m_l} \qquad (5.30)$$

This important equation will now be used to calculate the energies. It

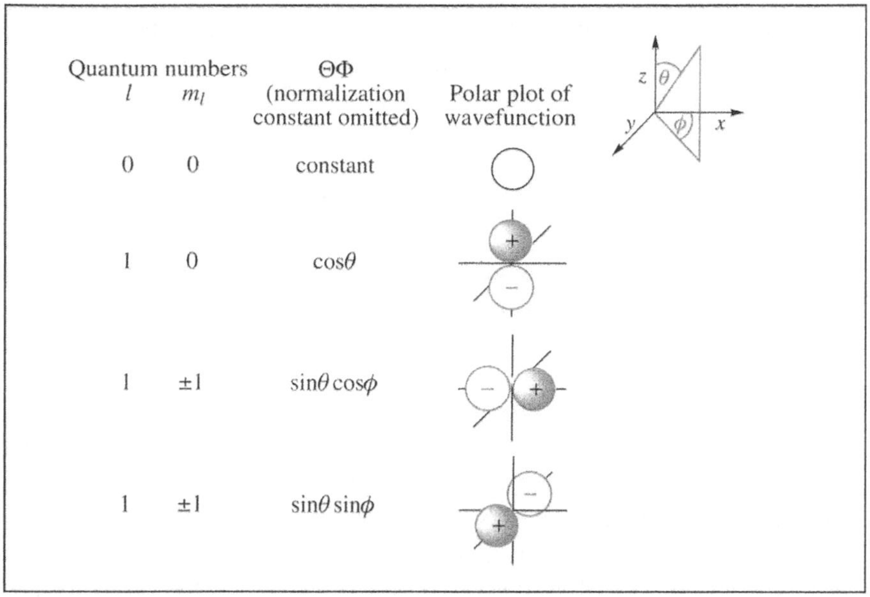

Quantum numbers		ΘΦ (normalization constant omitted)	Polar plot of wavefunction
l	m_l		
0	0	constant	
1	0	$\cos\theta$	
1	±1	$\sin\theta\cos\phi$	
1	±1	$\sin\theta\sin\phi$	

Figure 5.9 Polar plots of the first four spherical harmonic wavefunctions

will also appear in the derivation of the radial wavefunctions for the hydrogen atom.

A combination of equations (5.24) and (5.30) gives:

$$-\frac{\hbar^2}{2mr^2}\Lambda^2 Y_{l,m_l} = \frac{\hbar^2}{2mr^2}l(l+1)Y_{l,m_l} = EY_{l,m_l} \qquad (5.31)$$

The function Y_{l,m_l} appears on both sides of this equation and cancels out. From this it follows that the energy of a rotating particle is quantized according to the equation:

$$E = l(l+1)\frac{\hbar^2}{2I} \quad l = 0,\ 1,\ 2,\ 3,\ ... \qquad (5.32)$$

Because the energy is independent of the value of m_l, there will be $2l + 1$ states with the same energy, and the energy level is said to be $(2l + 1)$-fold degenerate

5.2.4 Angular Momentum and Spatial Quantization

The equation relating energy and angular momentum is $E = L^2/2I$ (see equation 5.5). It follows that the angular momentum is quantized, and limited to the following values:

$$L = \sqrt{l(l+1)}\hbar \quad l = 0,\ 1,\ 2,\ 3,\ ... \qquad (5.33)$$

It should be noted that the function $\Phi(\phi)$, which describes the motion of

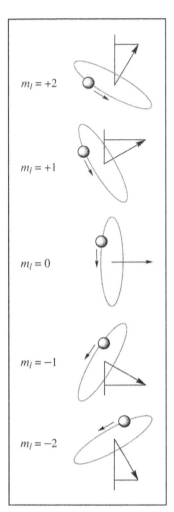

$m_l = +2$

$m_l = +1$

$m_l = 0$

$m_l = -1$

$m_l = -2$

Figure 5.10 The five directions in which the angular momentum vector can point for $l = 2$

the particle in the xy plane, is identical to that for a particle rotating in a fixed plane, for which the angular momentum was found to be equal to $m_l\hbar$. The component of angular momentum directed along the z-axis must therefore be quantized according to the equation:

$$L_z = m_l\hbar \quad m_l = 0, \ \pm1, \ \pm2, \ ... \ \pm l \tag{5.34}$$

These results can be represented by a vector with a length proportional to $\sqrt{l(l+1)}\hbar$, and oriented so that the projection of the vector on the z-axis is equal to $m_l\hbar$. This is illustrated in Figure 5.10.

For a given value of the quantum number l, there are $(2l + 1)$ directions in which the angular momentum vector can point. If all directions in space are equivalent, the position of the z axis is quite arbitrary, and the $2l + 1$ orientations that the rotating particle can adopt all have the same energy. In the presence of an externally applied electric or magnetic field, however, the z axis is determined by the direction of the field, and the orientation of the angular momentum with respect to this axis will affect the energy.

So far we have discussed the component of the angular momentum in the z direction, but no mention has been made of the components in the x and y directions. The reason for this omission is that the Uncertainty Principle forbids complete knowledge of the orientation of the angular momentum vector. If the component in the z direction is known, then the other two components must remain undetermined. This situation can be represented on a diagram by cones of uncertainty, as shown in Figure 5.11. One end of the angular momentum vector is

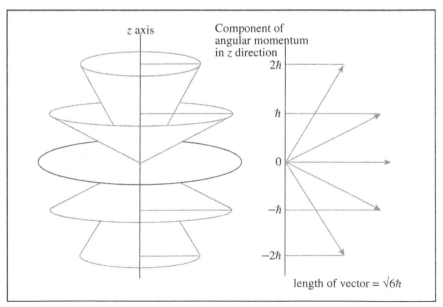

z axis

Component of angular momentum in z direction

$2\hbar$

$\hbar$

0

$-\hbar$

$-2\hbar$

length of vector $= \sqrt{6}\hbar$

Figure 5.11 The cones of uncertainty on which the angular momentum vectors must lie

considered to be at the apex of the appropriate cone and the other end can then be situated anywhere along the circular cross section at the top of the cone. This concept will be useful when we come to consider electron spins in a later section.

5.2.5 Application to the Rotation of Diatomic Molecules

The rotation of a rigid diatomic molecule about its centre of mass is illustrated in Figure 5.12a. As discussed in Chapter 4, the motion of the two masses, m_1 and m_2, is mathematically equivalent to the rotation of a single particle of mass μ [$= m_1 m_2 / (m_1 + m_2)$] about a fixed point, the distance between the point and the particle being equal to the bond length, r. This is illustrated in Figure 5.12b. It is therefore possible to treat the rotation of a diatomic molecule as the motion of a particle of mass μ on the surface of a sphere.

In rotational spectroscopy it is customary to use J as the quantum number, rather than l. With this modification to equation (5.32), the allowed energies of rotation can be written as:

$$E = \frac{\hbar^2}{2I} J(J+1) \quad J = 0, 1, 2 \dots \tag{5.35}$$

where the moment of inertia $I = \mu r^2$.

For a given rotational energy there will be $(2J + 1)$ possible orientations of the rotating molecule in space, and therefore each rotational state will be $(2J + 1)$-fold degenerate. Absorption of radiation in the far-infrared region of the spectrum causes molecules to become rotationally excited, the selection rule being that the quantum number J can only increase by one. Thus, the change in rotational energy resulting from the absorption of radiation can be obtained from the equation:

$$\Delta E(J \rightarrow J+1) = \frac{\hbar^2}{2I}\left[(J+1)(J+2) - J(J+1)\right] = \frac{\hbar^2}{I}(J+1) \tag{5.36}$$

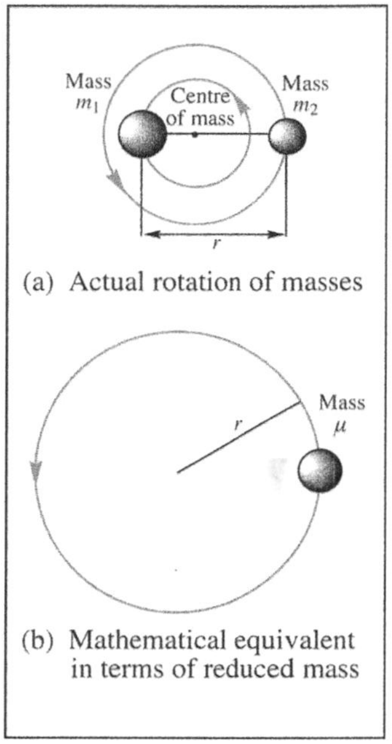

(a) Actual rotation of masses

(b) Mathematical equivalent in terms of reduced mass

Figure 5.12 The rotation of a diatomic molecule: (a) the actual situation; (b) the equivalent mathematical model used in the calculation

---

Worked Problem 5.4

Q In the far-infrared spectrum of $^{12}C^{16}O$ the rotational transition from $J = 12$ to $J = 13$ causes absorption of radiation at a wavenumber of 50.2 cm^{-1}. For the CO molecule calculate (a) the energy change involved in the transition, (b) the moment of inertia, (c) the reduced mass and (d) the bond length.

A (a) The energy of the photon absorbed in the transition can be calculated from the equation:

$$E\left(\text{photon}\right) = h\nu = \frac{hc}{\lambda}$$

With $1/\lambda = 5.02 \times 10^3 \text{ m}^{-1}$, this becomes:

$$E\left(\text{photon}\right) = \left(6.63 \times 10^{-34} \text{ J s}\right)\left(5.02 \times 10^3 \text{ m}^{-1}\right)\left(3.00 \times 10^8 \text{ m s}^{-1}\right)$$
$$= 9.98 \times 10^{-22} \text{ J}$$

(b) This energy is equal to the difference in rotational energy between the $J = 12$ and the $J = 13$ levels, as obtained from equation (5.36):

$$\Delta E\left(J = 12 \rightarrow J = 13\right) = \frac{13\hbar^2}{I}$$

Hence, the moment of inertia can be calculated as follows:

$$I = \frac{13 \times \left(1.054 \times 10^{-34} \text{ J s}\right)^2}{9.98 \times 10^{-22} \text{ J}} = 1.45 \times 10^{-46} \text{ kg m}^2$$

(c) The reduced mass is obtained from the equation

$$\mu = \left[\left(\frac{12.0 \times 16.0}{12.0 + 16.0}\right) \times 10^{-3} \text{ kg mol}^{-1}\right] \times \left[\frac{1}{6.02 \times 10^{23} \text{ mol}^{-1}}\right]$$
$$= 1.14 \times 10^{-26} \text{ kg}$$

(d) Finally, the bond length is obtained from the formula $I = \mu r^2$. Thus:

$$r = \sqrt{\frac{1.45 \times 10^{-46} \text{ kg m}^2}{1.14 \times 10^{-26} \text{ kg}}} = 1.13 \times 10^{-10} \text{ m}$$

5.3   Spin

Besides the angular momentum resulting from circular motion about a fixed point, many elementary particles have an intrinsic angular momentum which can be considered to arise from the particle spinning on its axis. This spin angular momentum is quantized in much the same way as the angular momentum of a particle moving over the surface of a sphere.

5.3.1 Electron Spin

The first experiment to demonstrate electron spin was carried out by Otto Stern and Walther Gerlach in 1921, and is illustrated in Figure 5.13. Silver atoms, formed by evaporation from a hot metal source in vacuum, were passed between the poles of an inhomogeneous magnet, and then condensed on to a glass plate. It was found that approximately half of the silver atoms were deflected upwards and half downwards, which resulted in two elongated spots on the detector plate. This observation could be explained only if the silver atoms were behaving like miniature magnets that were able to take up one of two orientations in the magnetic field, as shown in Figure 5.14. Atoms aligned with their south pole uppermost would be deflected upwards because the south pole would be in a stronger magnetic field than the north pole. By the same reasoning, silver atoms with the opposite orientation would be deflected downwards. The magnetic field had to be inhomogeneous, otherwise the forces acting on the north and south poles of the atoms would cancel one another out, and there would be no net deflection.

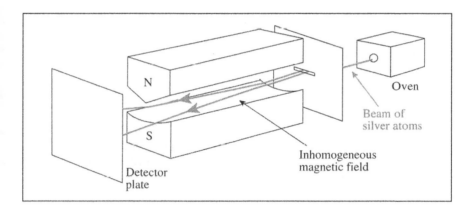

Figure 5.13 The Stern–Gerlach experiment to demonstrate the spin of the electron

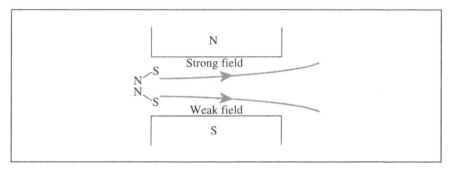

Figure 5.14 The deflection of the silver atoms in the inhomogeneous magnetic field

At the time when the experiments were performed the origin of the magnetic moment on the silver atom was uncertain. It was known from classical physics that an electron undergoing circular motion about an atomic nucleus would have a magnetic moment, but this could not explain the magnetic moment possessed by the silver atom because the single valence electron on the silver atom is in an s orbital, which has no orbital angular momentum. The reason for this will be discussed at length in Chapter 6. Later, it was suggested that the magnetic moment might arise because the valence electron was spinning on its axis, as illustrated in Figure 5.15. It should be emphasized that this is no more than a mental picture of a quantum mechanical phenomenon which has no true classical analogue, but it is a very useful model, provided it is not taken too far.

A spinning particle can be treated mathematically in much the same way as a particle moving on the surface of a sphere. The spin angular momentum vector is denoted by S and, by analogy with equation (5.33), we expect its magnitude to be given by the equation:

$$S = \hbar\sqrt{s(s+1)} \tag{5.37}$$

where s is the spin angular momentum quantum number (equivalent to l).

The projection of the spin angular momentum vector on to the z axis is denoted as S_z, and, by analogy with equation (5.34), it is restricted to multiples of $\hbar$ given by the equation:

$$S_z = m_s\hbar \qquad m_s = s, s-1, s-2, \dots -s \tag{5.38}$$

where m_s is another quantum number (equivalent of m_l). Thus, for a given value of the quantum number s, we expect there to be $2s + 1$ possible orientations of the spin angular momentum vector in the magnetic field. The experiments of Stern and Gerlach show that there are only two orientations that the spinning electron can take up. This requires that s be equal to $\frac{1}{2}$ and that $m_s = \pm\frac{1}{2}$. This is a departure from the conditions for orbiting particles, where the quantum number l was restricted to integral values. The boundary conditions for the electron spin wavefunctions are somewhat different from those for the motion of a particle on the surface of a sphere, and the quantum numbers are therefore subject to different rules. Generalizing from the work of Stern and Gerlach, it can be concluded that all electrons have an intrinsic spin angular momentum, given by the equation:

$$S = \hbar\sqrt{\tfrac{1}{2}\left(\tfrac{1}{2}+1\right)} = 0.866\hbar \tag{5.39}$$

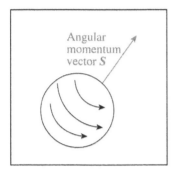

Figure 5.15 A picture of the electron spinning on its axis (not to be taken too literally)

Angular momentum vector S

Table 5.1 compares the quantization of angular momenta for both spin and orbital motion.

Table 5.1 Comparison of spin and orbital motion

Orbital motion	Spin
Orbital angular momentum,	Spin angular momentum,
$L = \hbar\sqrt{l(l+1)}$	$S = \hbar\sqrt{s(s+1)}$
Quantum number l = 0, 1, 2, ...	Quantum number $s = \frac{1}{2}$
Component of L in the z direction is restricted as follows: $L_z = m_l\hbar$	Component of S in the z direction is restricted as follows: $S_z = m_s\hbar$
Quantum number m_l = 0, ±1, ±2, ... ±l	Quantum number $m_s = ±\frac{1}{2}$
There are $2l + 1$ possible orientations of the orbital angular momentum for each value of l	There are only two possible orientations for electron spin

Worked Problem 5.5

Q For the Stern–Gerlach experiment, calculate the two angles that can be adopted by the spin angular momentum vector of the silver s electron with respect to the direction of the magnetic field.

A The two possible orientations that can be taken up by the spin angular momentum vector are shown in Figure 5.16. For the case where the spin is partially oriented in the direction of the field, the angle θ can be obtained from the equation:

$$\cos\theta = \frac{0.5\hbar}{0.866\hbar} = 0.577$$

Hence $\theta = 54.7°$. When the spin is partially opposed to the field, $\theta = 180 - 54.7 = 125.3°$. Although the angular momentum vector takes up a definite angle with respect to the z axis, its position with respect to the x and y axes remains undefined, and this is represented by the cones of uncertainty in the diagram.

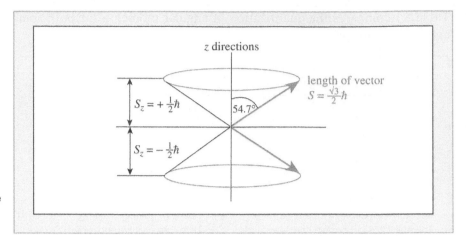

Figure 5.16 The two orientations that can be taken up by the spinning electron

5.3.2 Nuclear Spin of the Proton

Atomic nuclei also have spin, which is quantized in both magnitude and orientation in space. The nuclear spin quantum number is denoted by I, and the magnitude of the angular momentum is equal to $\hbar[I(I + 1)]^{1/2}$.

For the proton, $I = \frac{1}{2}$ and its angular momentum is therefore equal to $0.866\hbar$, which is the same value as that of the electron. However, because the mass of the proton is very much greater than that of the electron, the proton would need to spin much more slowly than the electron if the classical picture of the process is accepted.

In the presence of a magnetic field, the spin angular momentum vector of the proton can take up one of two orientations, just like the electron. This causes two energy levels to be produced, as shown in Figure 5.17. Nuclear magnetic resonance occurs when $\Delta E = h\nu$, where ν is the frequency of the applied radiofrequency field.

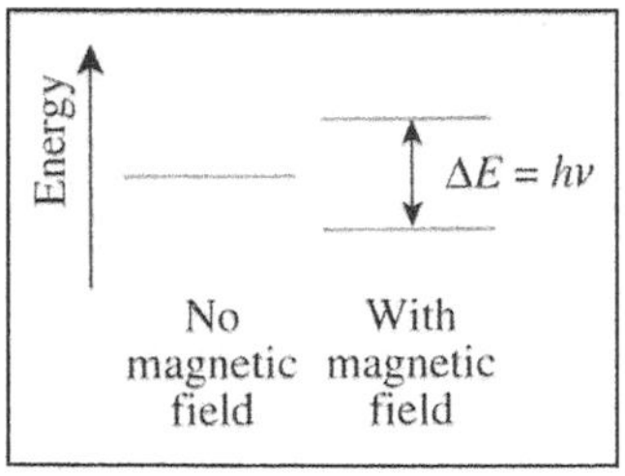

Figure 5.17 The two energy levels that are produced when a proton is placed in a magnetic field

5.3.3 Spins of Other Particles

Other elementary particles also have a characteristic spin. Those with half-integral spin quantum numbers are known as fermions and those with integral spin quantum numbers are known as bosons. The spin quantum numbers of a variety of particles are given in Table 5.2. When identical particles are interchanged, the wavefunctions associated with fermions and bosons behave differently, and this causes them to have significantly different properties. This will be discussed briefly in Chapter 7, where it is relevant to the way in which electron energy levels in atoms are filled.

Table 5.2 Spin quantum numbers, I, of some fundamental particles

Fermions		Bosons	
Particle	Spin quantum number, I	Particle	Spin quantum number, I
Electron	1/2	Photon	1
Proton (^{1}H)	1/2	Deuteron (^{2}H)	1
Neutron	1/2	^{4}He	0
^{13}C	1/2	^{12}C	0
^{35}Cl, ^{37}Cl	3/2	^{16}O	0

Appendix: Mathematics of Three-dimensional Rotational Motion

The Schrödinger equation for the motion of a particle on the surface of a sphere is:

$$-\frac{\hbar^2}{2I}\Lambda^2\psi = E\psi \qquad (A5.1)$$

where:

$$\Lambda^2 = \frac{1}{\sin\theta}\frac{\partial}{\partial\theta}\left(\sin\theta\frac{\partial\psi}{\partial\theta}\right) + \frac{1}{\sin^2\theta}\frac{\partial^2\psi}{\partial\phi^2} \qquad (A5.2)$$

We are looking for solutions of the form:

$$\psi = \Theta(\theta)\Phi(\phi) \qquad (A5.3)$$

When this expression for ψ is inserted into (A5.1), the equation becomes, after some rearrangement:

$$\frac{\Phi}{\sin\theta}\frac{d}{d\theta}\left(\sin\theta\frac{d\Theta}{d\theta}\right) + \frac{2IE}{\hbar^2}\Theta\Phi + \frac{\Theta}{\sin^2\theta}\frac{d^2\Phi}{d\phi^2} = 0 \qquad (A5.4)$$

Multiplication by $\sin^2\theta$ and division by $\Theta\Phi$ gives:

$$\frac{\sin\theta}{\Theta}\frac{d}{d\theta}\left(\sin\theta\frac{d\Theta}{d\theta}\right) + \frac{2IE}{\hbar^2}\sin^2\theta + \frac{1}{\Phi}\frac{d^2\Phi}{d\phi^2} = 0 \qquad (A5.5)$$

The two variables, θ and ϕ, can now be separated by moving the term in ϕ to the right-hand side:

$$\frac{\sin\theta}{\Theta}\frac{d}{d\theta}\left(\sin\theta\frac{d\Theta}{d\theta}\right)+\frac{2IE}{\hbar^2}\sin^2\theta=-\frac{1}{\Phi}\frac{d^2\Phi}{d\phi^2} \qquad (A5.6)$$

Since θ and ϕ can be varied independently of one another, this equation can be satisfied only when both sides of the equation are equal to a constant. Thus:

$$-\frac{1}{\Phi}\frac{d^2\Phi}{d\phi^2}=\text{const.} \qquad (A5.7)$$

Assuming that the constant is positive, the solutions of this equation are:

$$\Phi=Ne^{i\alpha\phi} \qquad (A5.8)$$

where α and N are constants. The constant in equation (A5.7) is then equal to α^2.

The variable ϕ represents rotation in the xy plane and this type of motion has already been discussed in Section 5.1.2, where satisfactory wavefunctions were found to take the form:

$$\Phi=\frac{1}{\sqrt{2\pi}}e^{im_l\phi} \quad \text{with } m_l=0,\ \pm1,\ \pm2,\ ... \qquad (A5.9)$$

This enables α to be equated with m_l, and the constant in equation (A5.7) then becomes m_l^2. Equation (A5.6) can then be written as:

$$\frac{\sin\theta}{\Theta}\frac{d}{d\theta}\left(\sin\theta\frac{d\Theta}{d\theta}\right)+\frac{2IE}{\hbar^2}\sin^2\theta=m_l^2 \qquad (A5.10)$$

After division by $\sin^2\theta$ and some rearrangement, this equation becomes:

$$\frac{1}{\Theta\sin\theta}\frac{d}{d\theta}\left(\sin\theta\frac{d\Theta}{d\theta}\right)-\frac{m_l^2}{\sin^2\theta}=-\frac{2IE}{\hbar^2} \qquad (A5.11)$$

Putting $2IE/\hbar^2$ equal to β, and multiplying throughout by Θ, we obtain:

$$\frac{1}{\sin\theta}\frac{\mathrm{d}}{\mathrm{d}\theta}\left(\sin\theta\frac{\mathrm{d}\Theta}{\mathrm{d}\theta}\right) - \frac{\Theta m_l^2}{\sin^2\theta} + \beta\Theta = 0 \qquad \text{(A5.12)}$$

Satisfactory solutions of this equation are found only when $\beta = l(l+1)$, where l is a positive integer greater than or equal to the modulus of m_l. Thus, the quantum numbers l and m_l are restricted to the following values:

$$l = 0, 1, 2, 3, \dots \text{ with } |m_l| \le l.$$

Inserting $\beta = l(l+1)$ and $m_l^2 = -\dfrac{1}{\Phi}\dfrac{\mathrm{d}^2\Phi}{\mathrm{d}\phi^2}$ into equation (A5.12) and rearranging, we obtain:

$$\frac{1}{\sin\theta}\frac{\mathrm{d}}{\mathrm{d}\theta}\left(\sin\theta\frac{\mathrm{d}\Theta}{\mathrm{d}\theta}\right) + \frac{\Theta}{\sin^2\theta}\frac{1}{\Phi}\frac{\mathrm{d}^2\Phi}{\mathrm{d}\phi^2} = -l(l+1)\Theta \qquad \text{(A5.13)}$$

Multiplying throughout by Φ gives:

$$\frac{\Phi}{\sin\theta}\frac{\mathrm{d}}{\mathrm{d}\theta}\left(\sin\theta\frac{\mathrm{d}\Theta}{\mathrm{d}\theta}\right) + \frac{\Theta}{\sin^2\theta}\frac{\mathrm{d}^2\Phi}{\mathrm{d}\phi^2} = -l(l+1)\Theta\Phi \qquad \text{(A5.14)}$$

The left side of this equation is equivalent to $\Lambda^2(\Theta\Phi)$ and therefore:

$$\Lambda^2(\Theta\Phi) = -l(l+1)\Theta\Phi \qquad \text{(A5.15)}$$

The wavefunctions that satisfy equation (A5.1) are known as spherical harmonics, and they are written as $Y_{l,m_l}(\theta,\phi)$, where l and m_l are the quantum numbers that determine the particular form that the wavefunction takes. With this notation, equation (A5.15) becomes:

$$\Lambda^2 Y_{l,m_l}(\theta,\phi) = -l(l+1)Y_{l,m_l}(\theta,\phi) \qquad \text{(A5.16)}$$

Summary of Key Points

1. The wavefunction for a particle moving in a circle has been shown to be:

$$\psi = \frac{1}{\sqrt{2\pi}}\,e^{im_l\phi}$$

where ϕ is the angle of rotation and m_l is a quantum number restricted to the values 0, ±1, ±2, *etc*. Positive and negative values of m_l represent motion in opposite directions around the circle. The angular momentum of the particle is equal to $m_l\hbar$, and the energy is given by the equation:

$$E = \frac{m_l^2\hbar^2}{2I}$$

where I is the moment of inertia of the particle.

2. The wavefunctions corresponding to the motion of a particle on the surface of a sphere are known as spherical harmonics. The mathematical form that they take depends upon two quantum numbers; the first is m_l, and the second is the azimuthal quantum number l. The rotational kinetic energy of the particle is quantized according to the equation:

$$E = l(l+1)\frac{\hbar^2}{2I}$$

The angular momentum, L, is equal to $\hbar[l(l+1)]^{1/2}$ and the component of angular momentum in any particular direction, L_z (taken to be the z direction), is equal to $m_l\hbar$. The angular momentum in the x and y directions is unknown. This leads to "cones of uncertainty", on which the angular momentum vector must lie.

3. The spherical harmonic wavefunctions can be applied to the rotational spectra of diatomic molecules, and also to electron and nuclear spin. In the latter case, the quantum numbers must be allowed to have half-integral values.

Problems

5.1. The wavefunction for a particle moving in a circle can be written as $\psi = N\sin\alpha\phi$, where N and α are constants and ϕ is the angle of rotation. Find the value of the normalization constant and the allowed values of the quantum number α. The following trigonometrical equation may be found useful:

$$\sin^2\theta = \tfrac{1}{2}\left(1 - \cos 2\theta\right)$$

5.2. The equilibrium internuclear distance in $^1H^{35}Cl$ is 0.1275 nm. Calculate the difference in rotational energy between the $J = 0$ and $J = 1$ levels and the wavelength of radiation that will be absorbed in promoting the transition from $J = 0$ to $J = 1$. The masses of ^1H and ^{35}Cl are 1.008 and 34.97 amu, respectively.

5.3. Draw vector diagrams to represent the angular momentum of states with the following quantum numbers: (a) $l = 2$, $m_l = 2$; (b) $l = 1$, $m_l = 0$; (c) $s = \tfrac{1}{2}$, $m_s = -\tfrac{1}{2}$.

5.4. A sample of deuterium is placed in a magnetic field. Calculate the minimum angle to the field that can be adopted by the spin angular momentum vector of the deuterium nucleus.

Further Reading

P. A. Cox, *Introduction to Quantum Theory and Atomic Structure*, Oxford University Press, Oxford, 1996.

J. W. Linnett, *Wave Mechanics and Valency*, Methuen, London, 1960.

G. M. Barrow, *Physical Chemistry*, 6th edn., McGraw-Hill, New York, 1996, chap. 9.

6

The Hydrogen Atom

Aims

By the end of this chapter you should be able to:

- Calculate the electron energy levels of the hydrogen atom from a knowledge of the Rydberg constant
- Understand the origins of the radial wave equation and the significance of the three energy terms that occur within it
- Explain the role of the three quantum numbers, n, l and m_l, in determining the overall structure and shape of the hydrogen atomic orbitals
- Demonstrate that the 1s wavefunction is indeed a solution of the Schrödinger equation for the hydrogen atom
- Understand the reasons why s-type wavefunctions have a maximum at the nucleus, whereas p- and d-type wavefunctions have a node there
- Calculate the distance from the nucleus at which a 1s electron is most likely to be found
- Understand the differences between the complex and real wavefunctions for p- and d-type wavefunctions

6.1 Introduction

In this chapter we shall consider atoms or ions which have a single electron. These include the hydrogen atom, He^+ and Li^{2+}. The solutions of Schrödinger's wave equation for such systems provide important information on the way in which an electron moves around the nucleus. The actual trajectory followed by an electron cannot be known in detail because of the operation of the uncertainty principle, but the wavefunctions obtained from the Schrödinger equation provide probability distributions for the position of the electron, and these are known as atomic orbitals. They can be contrasted with the accurately defined atomic orbits which would be required by classical physics.

The wavefunctions that we shall discuss form the basis of our understanding of atomic structure in general, because the concepts introduced can be extended to many-electron atoms. They will also prove useful when we come to discuss chemical bonding.

6.2 The Hydrogen Spectrum and the Quantization of Energy

The need for a quantum interpretation of the hydrogen atom arose from the spectrum of radiation emitted when an electric discharge was passed through hydrogen gas. In this process, electrons are promoted to higher energy states by the electric discharge, and photons are emitted when these electrons return to lower energy states. The energy released is given by the equation:

$$E_2 - E_1 = h\nu = \frac{hc}{\lambda} \tag{6.1}$$

where E_2 and E_1 are the energies of the upper and lower states, respectively. An electron moving in a classical orbit about the nucleus would be expected to emit radiation with a continuous range of wavelengths because there is no restriction on the energy that the orbiting electron can have. However, the actual spectrum consists of a series of lines with wavelengths that can be described by the formula:

$$\frac{1}{\lambda} = R\left(\frac{1}{n_1^2} - \frac{1}{n_2^2}\right) \tag{6.2}$$

In this equation, R is a constant, known as the Rydberg constant, and n_1 and n_2 are integers, with $n_2 > n_1$. Combining equations (6.1) and (6.2), we have:

$$E_2 - E_1 = hcR\left(\frac{1}{n_1^2} - \frac{1}{n_2^2}\right) \tag{6.3}$$

By associating E_2 with n_2 and E_1 with n_1 we can obtain the following equation for the allowed energies of the electron:

$$E_n = -\frac{hcR}{n^2} \tag{6.4}$$

where n must be an integer. The energies are negative because work has to be done to remove the electron from the region close to the nucleus

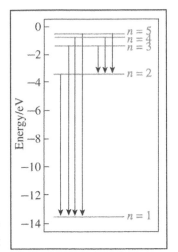

Figure 6.1 The energies of the allowed transitions in the spectrum of hydrogen

to infinity. The allowed energies and the transitions between them are illustrated in Figure 6.1.

Q Calculate the energy of the state corresponding to $n = 1$ in units of electronvolts, using $R = 1.097 \times 10^7$ m^{-1}.

A From equation (6.4):

$$E = -\left(6.626 \times 10^{-34} \text{ J s}\right) \times \left(2.998 \times 10^8 \text{ m s}^{-1}\right) \times \left(1.097 \times 10^7 \text{ m}^{-1}\right)$$

$$= 2.179 \times 10^{-18} \text{ J}$$

One electronvolt $= 1.602 \times 10^{-19}$ J; hence:

$$E = -\frac{2.179 \times 10^{-18} \text{ J}}{1.602 \times 10^{-19} \text{ J} \left(\text{eV}\right)^{-1}} = -13.60 \text{ eV}$$

This corresponds to the ionization energy of the hydrogen atom.

6.3   The Bohr Theory

In 1913, Bohr proposed a model for the hydrogen atom that appeared to explain the line spectra discussed in Section 6.2. The motion of the electron around the nucleus was considered to be similar to the motion of a planet around the sun, the gravitational attraction that keeps the planet in a circular or an elliptical orbit being replaced by the coulombic attraction between the electron and the positively charged nucleus. To account for the line spectra, Bohr postulated that the angular momentum of the electron was restricted to multiple values of $\hbar$. This was an arbitrary postulate at the time it was made, but it comes naturally from the quantum mechanical description of a particle moving in a circle, as we have already seen in Section 5.1.3.

Although the theory gives electron energies for the hydrogen atom which are in surprisingly good agreement with those calculated from equation (6.4), it fails to explain the energies and spectra obtained with other atoms, and it has now been completely superceded by quantum mechanics.

The Bohr theory has been mentioned here because the radius of the first Bohr orbit, known simply as the Bohr radius, is still widely used in quantum mechanics. It is given the symbol a_0 and has a value given by the formula:

$$a_0 = \frac{4\pi\varepsilon_0\hbar^2}{m_e e^2} = 52.9 \text{ pm} \tag{6.5}$$

In this equation, ε_0 is the permittivity of free space, m_e is the mass of the electron and e is the electronic charge.

6.4 Formulation of the Schrödinger Wave Equation for Hydrogen-like Atoms

These atoms consist of an electron and a nucleus, both of which are in motion. Since wave-like properties have to be associated with both particles, the full wave equation for the atom involves a total of six variables, and such equations are usually difficult to solve. Fortunately, the motion of the atom as a whole can be separated into two parts: (a) the translational motion of the centre of mass, for which particle-in-a-box wavefunctions are appropriate, and (b) the motion of the nucleus and electron relative to the centre of mass (see Figure 6.2a).

It can be shown[1] that the second form of motion is mathematically equivalent to the movement of a hypothetical particle, with reduced mass μ, about a fixed point. The reduced mass is given by the formula:

$$\mu = \frac{m_e m_n}{m_e + m_n} \tag{6.6}$$

where m_e and m_n are the masses of the electron and nucleus, respectively.

This is illustrated in Figure 6.2b. For hydrogen, the mass of the nucleus is 1836 times that of the electron, and the reduced mass is therefore very close to the mass of the electron. It follows that little error is involved if the electron is considered to be moving about a stationary nucleus, and this approximation is even better with other, heavier nuclei.

The positive charge on the nucleus is equal to $-Ze$, where Z is the atomic number of the atom, and the electron is attracted towards this charge with a force given by the inverse square law:

$$\text{force} = \frac{Ze^2}{4\pi\varepsilon_0 r^2} \tag{6.7}$$

r being the distance of the electron from the nucleus. The potential energy, V, of the electron is defined as the work done in bringing the electron from infinity to some specified point close to the nucleus, and it is obtained from the integral:

$$V = \int_{\infty}^{r} \frac{Ze^2}{4\pi\varepsilon_0 r^2} dr = -\frac{Ze^2}{4\pi\varepsilon_0 r} \tag{6.8}$$

Energy is actually released as the electron approaches the nucleus because the force is attractive, and V is therefore negative.

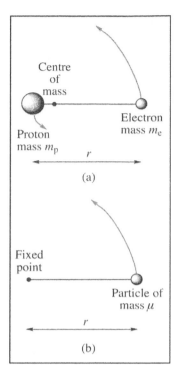

Figure 6.2 Diagram of the hydrogen atom: (a) the relative movements of the proton and electron about the centre of mass; (b) the mathematical equivalent of a particle of mass μ moving about a fixed point

The replacement of the relative motion of two particles with that of a hypothetical particle of mass μ, known as the **reduced mass**, is quite general, and has already been used in the description of the harmonic oscillator (Section 4.2.1) and the rotation of a diatomic molecule (Section 5.2.5).

In spherical polar coordinates the Schrödinger equation can now be written as:

$$-\frac{\hbar^2}{2\mu}\nabla^2\psi - \frac{Ze^2}{4\pi\varepsilon_0 r}\psi = E\psi \qquad (6.9)$$

where:

$$\nabla^2 = \frac{1}{r^2}\left[\frac{\partial}{\partial r}\left(r^2\frac{\partial}{\partial r}\right) + \Lambda^2\right] \qquad (6.10)$$

and

$$\Lambda^2 = \frac{1}{\sin\theta}\left[\frac{\partial}{\partial\theta}\left(\sin\theta\frac{\partial}{\partial\theta}\right)\right] + \frac{1}{\sin^2\theta}\frac{\partial^2}{\partial\phi^2} \qquad (6.11)$$

6.5 The Radial Wave Equation

Solutions of the Schrödinger equation can be found which are the product of three functions, each one involving only one of the variables r, θ and ϕ. The wavefunction can therefore be written as:

$$\psi = R(r)\Theta(\theta)\Phi(\phi) \qquad (6.12)$$

where the capital letters represent the functions and the lower case letters the associated variables. Substitution of this expression into equations (6.9) and (6.10) leads to:

$$-\frac{\hbar^2}{2\mu r^2}\left[\Theta\Phi\frac{\mathrm{d}}{\mathrm{d}r}\left(r^2\frac{\mathrm{d}R}{\mathrm{d}r}\right) + R\Lambda^2(\Theta\Phi)\right] - R\Theta\Phi\frac{Ze^2}{4\pi\varepsilon_0 r} = ER\Theta\Phi \quad (6.13)$$

Here, the variables upon which the functions operate have been omitted to reduce the length of the equation. The functions $\Theta(\theta)$ and $\Phi(\phi)$ are found to be exactly the same as the wavefunctions which were discussed in Chapter 5 for a particle on the surface of a sphere. These are the spherical harmonics, $Y_{l,m_l}(\theta, \phi)$, which depend upon the two quantum numbers, l and m_l. Thus, $\Theta(\theta)\Phi(\phi) = Y_{l,m_l}(\theta, \phi)$. It was shown in Chapter 5 that the spherical harmonics are characterized by the equation:

$$\Lambda^2 Y_{l,m_l}(\theta,\phi) = -l(l+1)Y_{l,m_l}(\theta,\phi) \qquad (6.14)$$

Combining this equation with equation (6.13) and writing Y in place of $\Theta(\theta)\Phi(\phi)$, we obtain:

$$-\frac{\hbar^2}{2\mu r^2}\left[Y\frac{\mathrm{d}}{\mathrm{d}r}\left(r^2\frac{\mathrm{d}R}{\mathrm{d}r}\right) - l(l+1)RY\right] - \frac{Ze^2 RY}{4\pi\varepsilon_0 r} = ERY \qquad (6.15)$$

The radial equation for the hydrogen atom is then obtained by dividing throughout by RY:

$$-\frac{\hbar^2}{2\mu r^2 R}\frac{d}{dr}\left(r^2\frac{dR}{dr}\right)+\frac{\hbar^2}{2\mu r^2}l(l+1)-\frac{Ze^2}{4\pi\varepsilon_0 r}=E \qquad (6.16)$$

This equation is very important because, without doing a lot of mathematics, it can provide a valuable insight into the motion of the electron about the nucleus. The first term in the equation involves the curvature of the radial component of the wavefunction, d^2R/dr^2, and it represents the radial kinetic energy of the electron as it moves towards, or away from, the nucleus. The second term is identical to the right hand side of equation (5.32), and it therefore represents the rotational kinetic energy of the electron as it revolves around the nucleus. This term is also sometimes referred to as the centrifugal energy. The two forms of motion are illustrated in Figure 6.3. The third term is the coulombic potential energy arising from the electrical attraction between the electron and the nucleus. These three energy terms all vary with the distance of the electron from the nucleus, but their sum must be independent of r because it is equal to the total energy of the system, which is constant.

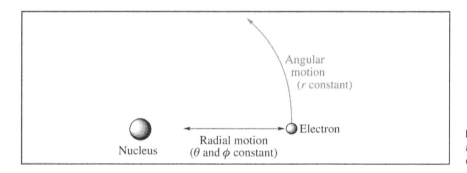

Figure 6.3 Illustration of the angular and radial motions of the electron

6.5.1 Solutions of the Radial Wave Equation

Equations similar to equation (6.16) have been studied by mathematicians, and acceptable solutions found. The mathematics involved are quite lengthy and will not be given here; they can be found in standard textbooks on quantum mechanics.[2] Two quantum numbers are needed to specify a particular radial wavefunction. The first one is the azimuthal quantum number, l, and the second one is a new quantum number n, often referred to as the principal quantum number. All the solutions have the same mathematical structure, which can be expressed by the equation:

$$R_{n,l}(r)=L_{n,l}(r)r^l e^{-Zr/na_0} \qquad (6.17)$$

Strictly, the quantity referred to as a_0 in equation (6.17) should be the Bohr radius multiplied by m_e/μ. Since $m_e/\mu = 1.0005$ for the hydrogen atom, the error involved in making this approximation is negligible. We shall use this approximation frequently from now on.

Here, $L_{n,l}(r)$ is a polynomial in r, and a_0 is the Bohr radius, already mentioned in Section 6.3.

The quantum number n can take the values $n = 1, 2, 3, \ldots$ and the quantum number l must always be less than n. Wavefunctions with $l = 0$, 1 and 2 are known as s, p and d orbitals, respectively. Thus, the wavefunction with $n = 2$ and $l = 1$ would be known as a 2p orbital.

The first few radial wavefunctions for hydrogen are listed in Table 6.1 and the way in which R and R^2 vary with distance from the nucleus is shown in Figure 6.4. The number of nodes (points where the wavefunction crosses the r axis) is equal to $n - l - 1$. The radial kinetic energy is related to the curvature of the wavefunction and increases with the number of nodes.

Table 6.1 The hydrogen-like radial wavefunctions $R(r)$ with $\sigma = Zr/a_0$, where
$$a_0 = \frac{4\pi\varepsilon_0\hbar^2}{\mu e^2}$$

Orbital type	n	l	$R_{nl}(r)$ (not normalized)
1s	1	0	$e^{-\sigma}$
2s	2	0	$(2 - \sigma)e^{-\sigma/2}$
2p	2	1	$\sigma e^{-\sigma/2}$
3s	3	0	$(27 - 18\sigma + 2\sigma^2)e^{-\sigma/3}$
3p	3	1	$(6 - \sigma)\sigma e^{-\sigma/3}$
3d	3	2	$\sigma^2 e^{-\sigma/3}$

Worked Problem 6.2

Q Use the radial wavefunctions for hydrogen, given in Table 6.1, to calculate the positions of the nodes for the 2s and 3s radial wavefunctions.

A The radial wavefunction, $R_{n,l}(r)$, goes to zero when the expression in front of the exponential term is equal to zero (see Table 6.1). For the 2s wavefunction, a node will occur when:

$$(2 - \sigma) = 0$$

where $\sigma = r/a_0$. Hence $r = 2a_0$.

For the 3s wavefunction we have to solve the quadratic equation:

$$27 - 18\sigma + 2\sigma^2 = 0$$

The roots of this equation are:

$$\sigma = \frac{18 \pm \sqrt{18^2 - (4 \times 2 \times 27)}}{4} = 1.90 \text{ and } 7.10$$

and nodes occur at $r = 1.90a_0$ and $7.10a_0$.

We see that the 3s wavefunction has its first node quite close to the node for the 2s wavefunction, but the second node is much further away from the nucleus.

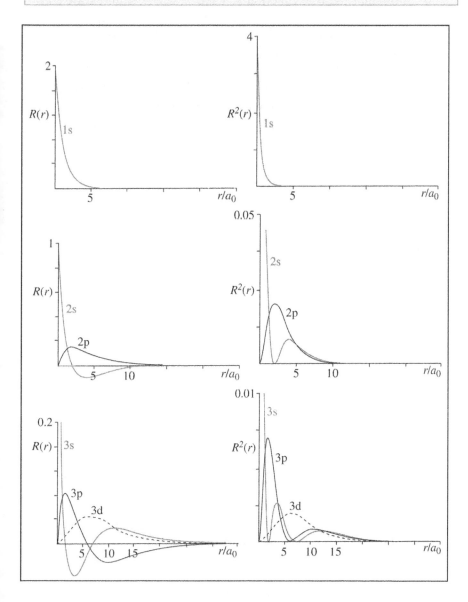

Figure 6.4 The variation of $R(r)$ and $R^2(r)$ with distance from the nucleus for the first few wavefunctions

6.5.2 A Relatively Simple Way to Find the Ground State Radial Wavefunction

One procedure for obtaining a solution of the radial wave equation is to make a guess as to the correct form of the wavefunction and see if it works. This is done in the example that follows.

Worked Problem 6.3

Q Show that the trial function $\psi = Ne^{-kr}$, where N and k are constants, is a solution of the Schrödinger equation for the hydrogen atom when the constant k has a particular value. Hence, calculate the energy associated with this wavefunction.

A Because the wavefunction does not depend upon the variables θ and ϕ, the spherical harmonic functions are not involved, and the operator Λ^2 can be omitted from equation (6.10). With $Z = 1$, the Schrödinger equation becomes:

$$-\frac{\hbar^2}{2\mu r^2}\left[\frac{\mathrm{d}}{\mathrm{d}r}\left(r^2\frac{\mathrm{d}\psi}{\mathrm{d}r}\right)\right] - \frac{e^2\psi}{4\pi\varepsilon_0 r} = E\psi \qquad (6.18)$$

This equation can also be obtained from equation (6.16) by putting $\psi = R$ and noting that the quantum number $l = 0$.

Differentiation of the trial wavefunction gives:

$$\frac{\mathrm{d}\psi}{\mathrm{d}r} = -kNe^{-kr} = -k\psi$$

Therefore:

$$\frac{\mathrm{d}}{\mathrm{d}r}\left(r^2\frac{\mathrm{d}\psi}{\mathrm{d}r}\right) = \frac{\mathrm{d}}{\mathrm{d}r}\left(-kr^2\psi\right) = \left(k^2r^2 - 2kr\right)\psi .$$

Equation (6.18) then becomes:

$$-\frac{\hbar^2}{2\mu}\left[k^2 - \frac{2k}{r}\right]\psi - \frac{e^2\psi}{4\pi\varepsilon_0 r} = E\psi \qquad (6.19)$$

Because E has to be constant, the terms in $1/r$ on the left-hand side must cancel one another out. This requires that:

$$\frac{\hbar^2 k}{\mu} = \frac{e^2}{4\pi\varepsilon_0}$$

It follows that:

$$k = \frac{\mu e^2}{4\pi\varepsilon_0 \hbar^2} \qquad (6.20)$$

A comparison with equation (6.5) shows that k is equal to the reciprocal of the Bohr radius a_0, provided we ignore the small difference between m_e and μ.

Thus, a satisfactory solution of the wave equation is:

$$\psi = N e^{-r/a_0} \qquad (6.21)$$

After the terms in $1/r$ have been removed from equation (6.19), we are left with:

$$E = -\frac{\hbar^2 k^2}{2\mu} = -\frac{\hbar^2}{2\mu}\left(\frac{\mu e^2}{4\pi\varepsilon_0 \hbar^2}\right)^2 = -\frac{\mu e^4}{32\left(\pi\varepsilon_0 \hbar\right)^2} \qquad (6.22)$$

As we shall see in Section 6.6.1, this is the energy of the ground state, with $n = 1$.

6.5.3 Behaviour of the Radial Wavefunction Close to the Nucleus

It can be seen from Figure 6.4 that R has a maximum value at the nucleus when $l = 0$, but is zero at the nucleus when $l = 1$ and $l = 2$. To understand this difference, we need to examine what happens to the energy terms in equation (6.16) as the electron approaches the nucleus and r tends to zero. If $l = 0$, the second term in equation (6.16) is zero, showing that there is no rotational energy and all the motion is along a radius. It can be seen from Figure 6.5 that the coulombic potential energy tends towards $-\infty$ as the electron approaches the nucleus, and the radial kinetic energy must therefore tend towards $+\infty$, in order to keep the total energy constant. For this to happen, dR/dr must be negative as $r \to 0$, which means that R must have a maximum value at the nucleus.

When $l > 0$, the situation is significantly different because the centrifugal energy varies as $1/r^2$, and becomes the dominant term close to the nucleus. This energy acts like a repulsive force and counteracts the

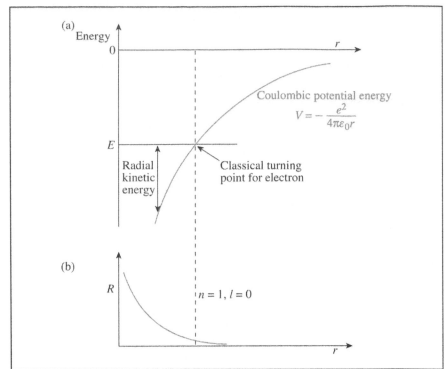

Figure 6.5 (a) The variation of the kinetic and coulombic potential energies of the electron with distance from the nucleus for $l = 0$, and (b) their effect on the radial wavefunction R. The dotted line represents the classical turning point for the electron; at greater distances the kinetic energy becomes negative and the electron can enter this region only by tunnelling

The radial kinetic energy of the electron also becomes negative at large distances from the nucleus where, owing to tunnelling, there is still a finite probability of finding it.

effect of the attractive coulomb potential. The variation of the centrifugal and potential energies with distance is shown in Figure 6.6, where it can be seen that the sum of the two energies becomes greater than the total energy, E, at distances less than a certain distance from the nucleus, r_c. This is a classically forbidden region, where the radial kinetic energy of the electron becomes negative. Although the electron can tunnel into this region (see Section 4.3.1), its wavefunction will decay away exponentially, and be zero at the nucleus.

6.6   The Full Hydrogen Atom Wavefunctions

The total wavefunction can be written as:

$$\psi = R_{n,l}(r)\Theta_{l,m_l}(\theta)e^{im_l\phi} \qquad (6.23)$$

The forms taken by this wavefunction for various values of the quantum numbers n, l and m_l (up to $n = 2$) are shown in Figure 6.7. The function $e^{im_l\phi}$ is the same one that we had for a particle moving around a circle, and here it represents the motion of the electron in the xy plane. The allowed values of the quantum number m_l are as follows:

$$m_l = 0, \pm1, \pm2, \pm3, \dots \text{ with } |m_l| \leq l$$

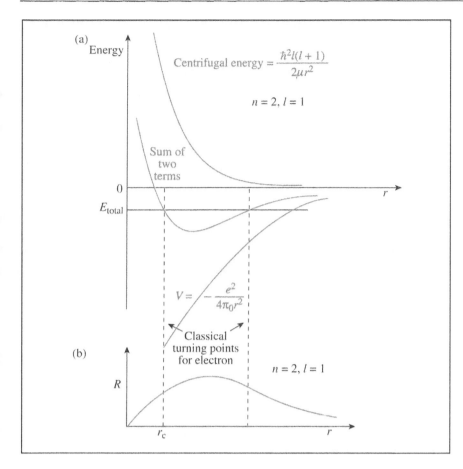

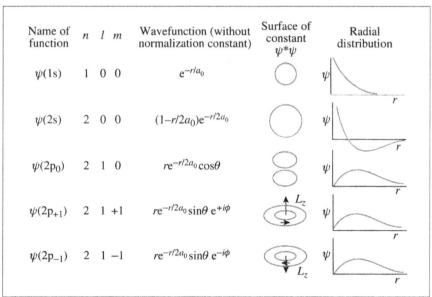

Figure 6.6 (a) The variation of the coulombic and centrifugal energies of the electron with distance from the nucleus for $l > 0$, and (b) their effect on the radial wavefunction R. There are two classical turning points and outside these limits the wavefunction decays away exponentially

Figure 6.7 The mathematical form and shapes of the first five hydrogen orbitals

The form taken by the θ function depends upon the two quantum numbers, l and m_l.

When $l = 0$, m_l must also equal zero, and the θ function becomes a constant. Thus, there is no angular variation to the wavefunction, and we obtain an s orbital. When $l = 1$, m_l can have the values 0, ±1. For $m_l = 0$, $\Theta(\theta) = \cos\theta$ and we obtain a $2p_z$ orbital, but we do not get the familiar $2p_x$ and $2p_y$ orbitals when $m_l = \pm 1$. Instead, we obtain doughnut-shaped orbitals in which the electrons can be circulating in either a clockwise or an anti-clockwise direction, and the wavefunction now has a complex component.

6.6.1 The Energies

The energies are obtained by solving the Schrödinger equation (equation 6.9). It is found that only the principal quantum number n is involved in determining the energy, which is given by the equation:

$$E_n = \frac{-\mu Z^2 e^4}{32 n^2 \pi^2 \varepsilon_0^2 \hbar^2} \quad n = 1, 2, 3 \tag{6.24}$$

The mathematics involved in obtaining this equation are quite lengthy, and will not be discussed here. A relatively simple way of obtaining the energy when $n = 1$ and $Z = 1$ has already been given in Worked Problem 6.3.

6.6.2 Angular Momentum

The rotational motion of the electron around the nucleus is quantized in a similar way to that of a particle on the surface of a sphere, which was described in Chapter 5. The total angular momentum of the electron is equal to $\hbar[l(l + 1)]^{1/2}$, and the component in the z direction is equal to $m_l \hbar$.

6.6.3 The s Orbitals

Motion of the s Electron and Shape of the Orbital

We have already seen that an electron has no rotational energy when $l = 0$ because the second term in equation (6.16) is zero. It follows that an s electron must undergo an oscillatory motion in a straight line through the nucleus, similar to that of a harmonic oscillator. Despite this similarity, the two forms of motion have different spatial properties because all directions in space are equivalent for an s electron, and the spherical shape of the s orbital arises from the uncertainty in the orientation of the oscillating electron. This is illustrated in Figure 6.8.

Although the circular orbit proposed by Bohr for an electron in the ground state of the hydrogen atom appeared to have some success, we now see that the actual motion of the electron is quite different. An oscillatory type of motion was originally considered by Bohr, but rejected as unacceptable because it would involve the electron colliding with the nucleus. This problem does not arise in such an acute form in wave mechanics because the positions of both nucleus and electron are uncertain, and neither can be precisely located without causing a major perturbation of the system.

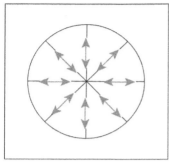

Figure 6.8 The movement of an electron in an s orbital

Probability Distributions

The most probable place to find an s electron is at the nucleus because the wavefunction has a maximum value there. Thus, if we move an imaginary electron detector of fixed volume, dV, around the atom, we will get a maximum reading at the nucleus. The probability depends upon the volume of the detector, and is equal to $\psi^*\psi dV$. The term $\psi^*\psi$ is equal to the probability per unit volume, and is known as the probability density (see Section 1.4.2). For the 1s orbital:

This is entirely a thought experiment because a real detector would perturb the electron to such an extent that no information about its original state could be obtained.

$$\psi_{1s}^* \psi_{1s} = N^2 e^{-2r/a_0} \qquad (6.25)$$

where N is a normalization constant. The variation of this expression with distance from the nucleus is shown in Figure 6.9a.

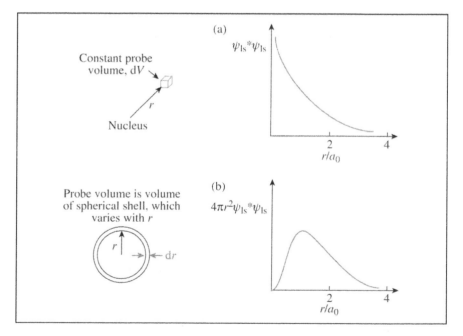

Figure 6.9 Two ways of recording the probability of an s electron being found at various distances from the nucleus: (a) using a constant volume probe; (b) using a spherical shell of variable volume

A different type of probability distribution can be obtained by asking the question: "what is the probability of finding the electron at a distance between r and $r + dr$ from the nucleus?" In this case the volume element that we have to consider (see Figure 6.9b) gets bigger as we move away from the nucleus. The volume of the spherical shell, dV, is now equal to $4\pi r^2 dr$, and the probability of finding the electron between r and $r + dr$ is equal to $4\pi r^2 \psi^* \psi dr$.

The volume of the spherical shell can be obtained by differentiating the equation for the volume of a sphere, $V = (4/3)\pi r^3$. Thus, $dV/dr = 4\pi r^2$.

Using the expresson for $\psi_{1s}^* \psi_{1s}$ from equation (6.25), the probability per unit radial length, P_r, becomes:

$$P_r = 4\pi r^2 N^2 e^{-2r/a_0} \tag{6.26}$$

This function is plotted in Figure 6.9b, where it can be seen that the probability goes through a maximum value at a distance from the nucleus equal to the Bohr radius, a_0. Close to the nucleus the r^2 term is dominant, which causes the probability to increase with r, but at distances greater than the Bohr radius the exponential term becomes dominant, and the probability falls with increasing r. It is surprising to find that the electron is most likely to be found at the Bohr radius, bearing in mind the large differences between the Bohr model and the wave mechanical one.

Worked Problem 6.4

Q Calculate the distance from the nucleus at which the 1s electron is most likely to be found.

A The maximum value of P_r can be obtained by putting dP_r/dr equal to zero. Hence:

Differentiation of equation (6.26) with respect to the variable r is carried out using the product rule: $\dfrac{d(uv)}{dr} = u\dfrac{\partial v}{\partial r} + v\dfrac{\partial u}{\partial r}$, where u and v are functions of r.

$$\frac{dP_r}{dr} = 4\pi N^2 \left(2r - \frac{2r^2}{a_0} \right) e^{-2r/a_0} = 0$$

Ignoring the exponential term, which goes to zero as $r \to \infty$, there are two solutions: $r = 0$ and $r = a_0$. The first solution corresponds to a minimum and the second to a maximum. Thus, the electron is most likely to found at $r = a_0$.

Normalization

The wavefunction for the 1s orbital can be written as:

$$\psi_{1s} = Ne^{-r/a_0} \tag{6.27}$$

where N is a normalization constant.

Worked Problem 6.5

Q Calculate the normalization constant for the hydrogen 1s orbital.

A We need to integrate P_r in equation (6.26) over all values of r, and equate the result to one. Thus:

$$\int_{r=0}^{r=\infty} 4\pi r^2 N^2 e^{-2r/a_0} dr = 4\pi N^2 \int_{r=0}^{r=\infty} r^2 e^{-2r/a_0} dr = 1$$

The integral is a standard form that can be looked up in mathematics textbooks:

$$\int_{x=0}^{x=\infty} x^n e^{-bx} dx = \frac{n!}{b^{n+1}}$$

Here, n is an integer and b is a constant. Replacing the variable x with r, and putting $n = 2$, $b = 2/a_0$, we find that the integral is equal to $(a_0^3)/4$. Thus:

$$4\pi N^2 \left(\frac{a_0^3}{4} \right) = 1$$

and:

$$N = \frac{1}{\sqrt{\pi a_0^3}} \qquad (6.28)$$

6.6.4 The p Orbitals

We have already seen that the $2p_0$ orbital, shown in Figure 6.7, is the same as the familiar $2p_z$ orbital, but the $2p_{\pm 1}$ orbitals involve complex wavefunctions that are not so easy to visualize. For these orbitals, the electron is circulating around the nucleus in the xy plane, and has a component of angular momentum in the z direction equal to $\pm \hbar$.

The more familiar $2p_x$ and $2p_y$ orbitals take the forms:

$$\psi_{2p_x} = N_x x \, e^{-r/2a_0} \qquad (6.29)$$

and

$$\psi_{2p_y} = N_y y \, e^{-r/2a_0} \qquad (6.30)$$

where N_x and N_y are normalization constants, and x and y are Cartesian

Only wavefunctions with the
same energy can be combined in
this way to give another
acceptable wavefunction. See
Section 5.2.2.

coordinates. These functions can be obtained by a linear combination of the complex functions, as shown in the Appendix at the end of the chapter.

This procedure gives:

$$\psi_{2p_{+1}} + \psi_{2p_{-1}} = \sqrt{2}\,\psi_{2p_x} \tag{6.31}$$

and

$$\psi_{2p_{+1}} - \psi_{2p_{-1}} = \sqrt{2}\,\psi_{2p_y} \tag{6.32}$$

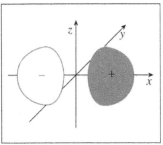

Figure 6.10 Surface of constant ψ for an electron in a $2p_x$ orbital

The shape of the $2p_x$ orbital can be deduced from equation (6.29). The presence of the Cartesian coordinate x in the expression for $2p_x$ means that there will be a node in the wavefunction in the yz plane. The surface of constant ψ is illustrated in Figure 6.10. The full three-dimensional orbital shape can be obtained by rotating the profile about the x axis. It is seen to consist of two doughnut-shaped lobes of opposite sign. The $2p_y$ orbital will be similar, but oriented along the y axis.

The $2p_x$ and $2p_y$ orbitals are not associated with any particular value of the angular momentum because they have been constructed from orbitals in which the electrons are moving in opposite directions. In this respect they are inferior to the $2p_{\pm 1}$ orbitals because some information has been lost, but their directional properties make them very important in chemical bonding.

6.6.5 The d Orbitals

These wavefunctions occur when n $\geq$ 3 and $l = 2$. For each value of n there are five distinct orbitals, corresponding to $m_l = 2, 1, 0, -1, -2$. Each of the orbitals has a total angular momentum equal to $\sqrt{6}\hbar$, but they have different orientations, the component of angular momentum along the z axis being equal to $m_l\hbar$. The situation is similar to that shown in Figure 5.11 for the spherical harmonics.

Four of these wavefunctions are complex, because they include the term $e^{im_l\phi}$, but real wavefunctions can be obtained by combining pairs of complex wavefunctions with the same $|m_l|$, but opposite sign. The orbitals obtained by this process are listed in Table 6.2, and their shapes are shown in Figure 6.11. Linear combinations of the complex orbitals corresponding to $m_l = \pm 1$ give the $3d_{xz}$ and $3d_{yz}$ orbitals, whereas those with $m_l = \pm 2$ combine to give the $3d_{x^2-y^2}$ and $3d_{xy}$ orbitals. The $3d_{z^2}$ orbital, corresponding to $m_l = 0$, does not have a complex component.

Table 6.2 The real 3d wavefunctions

n	l	m_l	Real wavefunction (not normalized)[a]
3	2	0	$\psi(3d_{z^2}) = \sigma^2 e^{-\sigma/3}(3\cos^2\theta - 1)$
3	2	± 1	$\psi(3d_{xz}) = \sigma^2 e^{-\sigma/3}\sin\theta\cos\theta\cos\phi$
			$\psi(3d_{yz}) = \sigma^2 e^{-\sigma/3}\sin\theta\cos\theta\sin\phi$
3	2	± 2	$\psi(3d_{x^2-y^2}) = \sigma^2 e^{-\sigma/3}\sin^2\theta\cos2\phi$
			$\psi(3d_{xy}) = \sigma^2 e^{-\sigma/3}\sin^2\theta\sin2\phi$

[a] $\sigma = Zr/a_0$

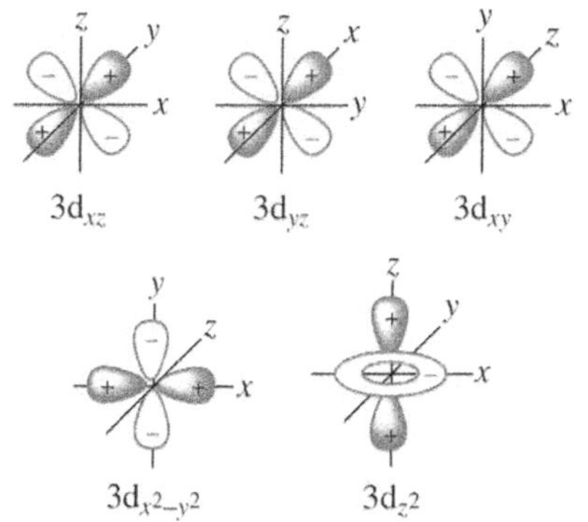

Figure 6.11 Surfaces of constant $\psi^*\psi$ for the 3d orbitals. The sign of the wavefunction is indicated, but the probability density itself will, of course, always be positive. Note that the orientation of the Cartesian axes is not the same in all the figures

Appendix: Formation of the 2p$_x$ and 2p$_y$ Orbitals from the Complex 2p$_{+1}$ and 2p$_{-1}$ Orbitals

The 2p$_x$ and 2p$_y$ orbitals can be obtained by a linear combination of the complex functions:

$$\psi_{2p_{+1}} \pm \psi_{2p_{-1}} = N_{\pm 1}\, r\, e^{-r/2a_0}\sin\theta\left(e^{+i\phi} \pm e^{-i\phi}\right) \qquad (A6.1)$$

where $N_{\pm 1}$ is the normalization constant for the complex orbitals. Now:

$$\begin{aligned}
e^{+i\phi} \pm e^{-i\phi} &= \left[\cos\phi + i\sin\phi\right] \pm \left[\cos(-\phi) + i\sin(-\phi)\right] \\
&= \left[\cos\phi + i\sin\phi\right] \pm \left[\cos\phi - i\sin\phi\right] \\
&= 2\cos\phi \ \ \text{or} \ \ 2i\sin\phi
\end{aligned} \qquad (A6.2)$$

Hence:

$$\psi_{2p_{+1}} + \psi_{2p_{-1}} = 2N_{\pm 1} r\, e^{-r/2a_0} \sin\theta \cos\phi \qquad (A6.3)$$

and

$$\psi_{2p_{+1}} - \psi_{2p_{-1}} = 2iN_{\pm 1} r\, e^{-r/2a_0} \sin\theta \sin\phi \qquad (A6.4)$$

Referring back to Figure 5.7, it can be seen that $r\sin\theta\cos\phi$ is equal to the Cartesian coordinate x, and $r\sin\theta\sin\phi$ is equal to the coordinate y. The wavefunctions can therefore be written as:

$$\psi_{2p_{+1}} + \psi_{2p_{-1}} = 2N_{\pm 1} x\, e^{-r/2a_0} \qquad (A6.5)$$

and

$$\psi_{2p_{+1}} - \psi_{2p_{-1}} = 2iN_{\pm 1} y\, e^{-r/2a_0} \qquad (A6.6)$$

The function on the right-hand side of equation (A6.5) represents a $2p_x$ orbital because it has a node in the yz plane. Similarly, the function on the right-hand side of equation (A6.6) represents a $2p_y$ orbital. As they stand, these functions are not normalized. After normalization, which involves integration over all possible values of r, θ and ϕ, we can write:

$$\psi_{2p_{+1}} + \psi_{2p_{-1}} = \sqrt{2}\,\psi_{2p_x} \qquad (A6.7)$$

and

$$\psi_{2p_{+1}} - \psi_{2p_{-1}} = \sqrt{2}\,\psi_{2p_y} \qquad (A6.8)$$

Note that the "i" in equation (A6.6) has now disappeared.

Summary of Key Points

1. The wavefunction for the hydrogen atom can be expressed as the product of three functions in the variables r, θ and ϕ:

$$\psi = R_{n,l}(r)\Theta_{l,m_l}(\theta)e^{im_l\phi}$$

The mathematical form taken by the wavefunction is dependent upon three quantum numbers n, l and m_l. These quantum numbers can take integral values, subject to the following restrictions:

$$n = 1, 2, 3, etc. \qquad l < n \qquad |m_l| \leq l$$

Orbitals with $l = 0$, 1 and 2 are known as s, p and d orbitals, respectively.

2. The motion of an electron in an s orbital is entirely along a radius, and the wavefunction has a maximum value at the nucleus. However, the electron is most likely to be found at a distance from the nucleus equal to the Bohr radius.

3. Electrons with $n = 2$, $l = 1$ and $m_l = 0$, +1 and –1 are said to occupy the $2p_0$, $2p_{+1}$ and $2p_{-1}$ orbitals, respectively. The $2p_0$ orbital is identical to the $2p_z$ orbital, but the other two orbitals are complex and differ from the more familiar $2p_x$ and $2p_y$ orbitals. The latter can be obtained by linear combinations of the two complex orbitals.

4. Electrons with $n = 3$, $l = 2$ and $m_l = 0$, ±1, ±2 are said to occupy the 3d orbitals. Four of these orbitals are complex, but real ones can be obtained by linear combinations of the complex orbitals.

Problems

6.1. Calculate the energy in eV required to excite an electron in a hydrogen atom from the $n = 1$ to the $n = 2$ state, given that the Rydberg constant is equal to 1.097×10^7 m^{-1}.

6.2. Calculate the distance from the nucleus at which the 2s radial wavefunction has a minimum value.

6.3. Calculate the distance from the nucleus at which the radial wavefunction for the 3p orbital has (a) a node, (b) a maximum value and (c) a minimum value. Hence sketch the 3p radial wavefunction.

6.4. At what distance from the nucleus is an electron in a hydrogen 2s orbital most likely to be found? *Hint*: there are two maxima in the radial probability curve, separated by a node at $r = 2a_0$. The second maximum is larger than the first and is therefore the one that is required.

6.5. For the 1s orbital of the hydrogen atom, calculate the average distance, $\langle r \rangle$, of the electron from the nucleus. *Hint*: $\langle r \rangle = \int_0^{\infty} rP(r)dr$, where $P(r)$ is the probability per unit length (see equation 6.26).

You will also need to use the standard integral: $\int_0^{\infty} x^n e^{-bx}dx = \dfrac{n!}{b^{n+1}}$, where n is an integer and b is an arbitrary constant.

6.6. Show that the radial wavefunction

$$R\left(r\right) = Nr\,e^{-r/2a_0}$$

where N is a normalization constant, is a solution of equation (6.16) when the quantum number $l = 1$. Hence, calculate the energy of this state.

6.7. At what distance along the z axis is an electron in a $2p_z$ orbital of a hydrogen atom most likely to be found? *Note*: this is not the same as the most probable radius. The correct answer will be obtained by considering a probe of constant volume, which is moved along the z axis.

6.8. In one of the excited states of the hydrogen atom, the electron has a total orbital angular momentum of $\sqrt{2}\hbar$ but no component of angular momentum along the z axis. The minimum energy required to remove the electron from the atom is 3.4 eV. Which orbital does the electron occupy?

6.9. The d orbitals, corresponding to the quantum numbers $n = 3$, $l = 2$ and $m_l = \pm 1$, have the following wavefunctions:

$$\psi = R\left(r\right)\sin\theta\cos\theta\,e^{\pm i\phi}$$

where $R(r)$ is the radial wavefunction. What real wavefunctions can be obtained from these complex wavefunctions?

References

1. P. W. Atkins, *Physical Chemistry*, 6th edn., Oxford University Press, Oxford, 1998, p. 346.
2. R. Eisberg and R. Resnick, *Quantum Physics of Atoms, Molecules, Solids, Nuclei, and Particles*, 2nd edn., Wiley, New York, 1985, appendix N.

Further Reading

D. A. McQuarrie and J. D. Simon, *Physical Chemistry: A Molecular Approach*, University Science Books, Sausalito, California, 1997, chap. 6.
P. W. Atkins and R. S. Friedman, *Molecular Quantum Mechanics*, 3rd edn., Oxford University Press, Oxford, 1997, chap. 7.

7

Further Concepts in Quantum Mechanics and their Application to Many-electron Atoms

Aims

By the end of this chapter you should be able to:

*   Write down the Hamiltonian operator for atoms containing two or more electrons
*   Explain the approximations used in the one-electron approximation, as applied to many-electron atoms
*   Write down the spin states of helium in its ground and first excited states
*   Explain how the penetration of outer electrons into inner shells affects the energies of s, p and d orbitals
*   Explain qualitatively how orbital energies can be calculated using the Hartree–Fock method
*   Use the Pauli exclusion principle and the Aufbau principle to find the electronic configuration of an element
*   Use Hund's rule to calculate the number of parallel electron spins in an atom
*   Explain the variation of the first ionization energy of atoms with atomic number

In previous chapters we have considered systems for which there is an exact solution to the Schrödinger wave equation, but as we begin to look at atoms containing more than one electron we shall find that it is impossible to solve the Schrödinger equation exactly, and various approximations will have to be introduced to make the problem solvable. Before these are considered it will be useful to look at some basic concepts of quantum mechanics in more detail, so that we can obtain the Schrödinger equation for any system that may be of interest.

7.1   The Hamiltonian Operator

The three-dimensional Schrödinger wave equation for a particle of mass m moving in a potential energy field V was written in Chapter 5 as:

$$-\frac{\hbar^2}{2m}\nabla^2\psi + V\psi = E\psi \qquad (7.1)$$

Here, $\nabla^2 \equiv \dfrac{\partial^2}{\partial x^2} + \dfrac{\partial^2}{\partial v^2} + \dfrac{\partial^2}{\partial z^2}$, and is known as the Laplacian operator.

The wavefunction, ψ, and the potential energy, V, are both functions of the coordinates of the particle, x, y and z. The total energy of the particle is represented by E, which must be constant.

This equation can be rearranged to give:

$$\left(-\frac{\hbar^2}{2m}\nabla^2 + V\right)\psi = E\psi \qquad (7.2)$$

The expression in parentheses is known as the Hamiltonian operator and is given the symbol $\hat{H}$. It contains instructions for the mathematical manipulation of whatever function follows it (for example, differentiate twice with respect to x, y and z). The cap over the "H" is a reminder that this term is an operator, and not simply a multiplier. With this shorthand notation the Schrödinger equation can be written simply as:

$$\hat{H}\psi = E\psi \qquad (7.3)$$

where $\hat{H} \equiv -\dfrac{\hbar^2}{2m}\nabla^2 + V$.

The Hamiltonian operator is named after Sir William Rowan Hamilton, an Irish mathematician, who devised an alternative form of Newton's equations of motion involving a function H, known as the Hamilton function. For most classical systems, H turns out to be simply the total energy of the system expressed in terms of the coordinates of the particles and their conjugate momenta. The kinetic energy of a single particle of mass m can be written as:

$$T = \frac{1}{2m}\left(p_x^2 + p_y^2 + p_z^2\right) \qquad (7.4)$$

where p_x, p_y and p_z are the components of linear momentum in the x, y and z directions, respectively. The potential energy, $V(x, y, z)$, is some unspecified function of the coordinates. The Hamiltonian function is then defined simply as the sum of the kinetic and potential energies:

$$H \equiv T + V = \frac{1}{2m}\left(p_x^2 + p_y^2 + p_z^2\right) + V\left(x, y, z\right) \qquad (7.5)$$

Thus, $H = E$, the total energy of the system.

The entirely classical Hamiltonian function, H, can be converted into the Hamiltonian operator, $\hat{H}$, by applying some simple rules, which can be stated as follows:

1. Obtain the classical equation for H in terms of the Cartesian co-ordinates of all the particles and their associated momenta, p_q (q stands for x, y or z).
2. Wherever a component of momentum, p_q, occurs, replace it with the operator $-i\hbar\dfrac{\partial}{\partial q}$.
3. Leave the positional coordinates unchanged.

These rules may seem very strange at first but, as we shall see, their application leads to the familiar Schrödinger equation.

7.2   Application to the Motion of a Single Particle

We can convert the Hamiltonian function for a single particle into the Hamiltonian operator by substituting the expressions for p_x, p_y and p_z into equation (7.5). This gives:.

$$\hat{H} = \frac{1}{2m}\left\{\left(-i\hbar\frac{\partial}{\partial x}\right)^2 + \left(-i\hbar\frac{\partial}{\partial y}\right)^2 + \left(-i\hbar\frac{\partial}{\partial z}\right)^2\right\} + V \qquad (7.6)$$

Because the potential energy is a function of the coordinates alone and does not involve the momenta, it remains unchanged in this process. After multiplying out the terms, the equation becomes:

$$\hat{H} = -\frac{\hbar^2}{2m}\left\{\frac{\partial^2}{\partial x^2} + \frac{\partial^2}{\partial y^2} + \frac{\partial^2}{\partial z^2}\right\} + V = -\frac{\hbar^2}{2m}\nabla^2 + V \qquad (7.7)$$

A comparison of this equation with equation (7.3) shows that this is indeed the Hamiltonian operator for a single particle, previously obtained from the Schrödinger equation.

7.3   Eigenfunctions and Eigenvalues

The Schrödinger equation, when written as $\hat{H}\psi = E\psi$, is an example of a special class of equations which can be put into the general form:

$$\hat{A}\phi = a\phi \tag{7.8}$$

Here, $\hat{A}$ is an operator which operates on the function ϕ to give the same function back again, multiplied by a constant "a". Any function which satisfies the equation is known as an eigenfunction of the operator $\hat{A}$ ("eigen" is the German word for "characteristic"), and the constant "a" is said to be an eigenvalue of the operator $\hat{A}$.

In general, there will be many functions ϕ which satisfy the equation, each with its own eigenvalue. If we indicate each separate solution by the index "n", the equation can be written as:

$$\hat{A}\phi_n = a_n\phi_n \tag{7.9}$$

When the operator is the Hamiltonian, we have:

$$\hat{H}\psi_n = E_n\psi_n \tag{7.10}$$

where the wavefunctions, ψ_n, are the eigenfunctions of the Hamiltonian operator and the allowed energies are the eigenvalues.

7.4   The Wave Equation for the Helium Atom

The approach outlined in Section 7.1 makes it possible to write down the Schrödinger equation for any system that we wish to study. To demonstrate the method we shall take the helium atom as an example. The distances between the nucleus and the two electrons are shown in Figure 7.1. In general, the charge on the nucleus will be $-Ze$, where Z is the atomic number of the element. For helium, $Z = 2$.

We start by writing down the classical equations for this system. If the components of linear momentum for electrons 1 and 2 are p_{x_1}, p_{y_1}, p_{z_1} and $p_{x_2}, p_{y_2}, p_{z_2}$, respectively, the kinetic energy can be written as:

$$T = \frac{1}{2m_e}\left(p_{x_1}^2 + p_{y_1}^2 + p_{z_1}^2\right) + \frac{1}{2m_e}\left(p_{x_2}^2 + p_{y_2}^2 + p_{z_2}^2\right) \tag{7.11}$$

The potential energy consists of three terms: the attractions of the two electrons to the nucleus, which make a negative contribution, and the inter-electron repulsion. Thus:

$$V = -\frac{e^2}{4\pi\varepsilon_0}\left(\frac{Z}{r_1} + \frac{Z}{r_2} - \frac{1}{r_{12}}\right) \tag{7.12}$$

The Hamiltonian function is the sum of these two terms: $H = T + V$.

The quantum mechanical equivalents are then obtained by making the

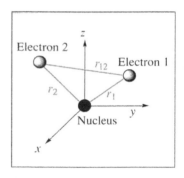

Figure 7.1 Schematic diagram of the helium atom, showing the distances between the nucleus and the two electrons

substitutions outlined in Section 7.1. These give:

$$\hat{H} = -\frac{\hbar^2}{2m_e}\left(\nabla_1^{\ 2} + \nabla_2^{\ 2}\right) + V \qquad (7.13)$$

where $\nabla_1^{\ 2} = \left(\dfrac{\partial^2}{\partial x_1^2} + \dfrac{\partial^2}{\partial y_1^2} + \dfrac{\partial^2}{\partial z_1^2}\right)$ and similarly for ∇_2^2.

The Schrödinger equation for the two electrons then becomes:

$$\left[-\frac{\hbar^2}{2m_e}\left(\nabla_1^{\ 2} + \nabla_2^{\ 2}\right) + V\right]\Psi = E\Psi \qquad (7.14)$$

where Ψ is the total wavefunction for the atom.

Two points are worthy of note:

(i) The physical significance of Ψ is slightly different from that for the wavefunction of a single particle. $\Psi^*\Psi$ now relates to the probability of simultaneously finding electron 1 at x_1, y_1, z_1 and electron 2 at x_2, y_2, z_2.

(ii) The total wavefunction, Ψ, is now a function of the coordinates of both electrons. It is virtually impossible to solve a second-order differential equation with six independent variables, and approximation methods must therefore be used to obtain the wavefunctions for this relatively simple system.

This problem can be greatly simplified if the electrical repulsion that exists between the two electrons is ignored. This is represented by the term $e^2/(4\pi\varepsilon_0 r_{12})$ in equation (7.12). By omitting this term we are treating the electrons as though they moved independently of one another, whereas in reality they tend to avoid one another because of the electrical repulsion.

This approximation allows each electron to be assigned its own hydrogen-like wavefunction, which is independent of the position of the other electron. These wavefunctions will be designated $\psi_{1s}(1)$ and $\psi_{1s}(2)$, where the number in parentheses identifies the electron occupying a particular orbital. The total wavefunction can then be written as:

$$\Psi(1,2) = \psi_{1s}(1)\psi_{1s}(2) \qquad (7.15)$$

This is known as the one-electron or orbital approximation because each electron is considered to occupy its own orbital, which is essentially hydrogen-like in character.

Worked Problem 7.1

Q Show that the wavefunction given in equation (7.15) is an eigenfunction of the Hamiltonian operator for the helium atom when the electron repulsion term is ignored.

A The simplified Hamiltonian operator can be divided into two parts, one relating to electron 1, and the other to electron 2:

$$\hat{H} = \hat{H}_1 + \hat{H}_2 \tag{7.16}$$

where

$$\hat{H}_1 = -\frac{\hbar^2}{2m_e}\nabla_1^2 - \frac{Ze^2}{4\pi\varepsilon_0 r_1} \tag{7.17}$$

and

$$\hat{H}_2 = -\frac{\hbar^2}{2m_e}\nabla_2^2 - \frac{Ze^2}{4\pi\varepsilon_0 r_2} \tag{7.18}$$

The one-electron wavefunctions are eigenfunctions of these two Hamiltonian operators:

$$\hat{H}_1\psi_{1s}(1) = E_1\psi_{1s}(1) \tag{7.19}$$

and

$$\hat{H}_2\psi_{1s}(2) = E_2\psi_{1s}(2) \tag{7.20}$$

The total wave equation then becomes:

$$\hat{H}\Psi(1,2) = (\hat{H}_1 + \hat{H}_2)\left[\psi_{1s}(1)\psi_{1s}(2)\right]$$

$$= \left[(\hat{H}_1\psi_{1s}(1))\psi_{1s}(2) + \psi_{1s}(1)[\hat{H}_2\psi_{1s}(2)]\right]$$

$$= \left[E_1\psi_{1s}(1)\right]\psi_{1s}(2) + \psi_{1s}(1)\left[E_2\psi_{1s}(2)\right]$$

$$= (E_1 + E_2)\psi_{1s}(1)\psi_{1s}(2)$$

$$= E\Psi(1,2) \tag{7.21}$$

Hence, the wavefunction $\psi_{1s}(1)\psi_{1s}(2)$ is an eigenfunction of the Hamiltonian operator $(\hat{H}_1 + \hat{H}_2)$.

The Hamiltonian operator $\hat{H}_2$ operates only on the coordinates of electron 2, and $\psi(1)$ is therefore treated as a constant when it follows this operator. Thus, in the derivation of equation (7.21) we obtain:

$$\hat{H}_2\left[\psi(1)\psi(2)\right] = \psi(1)\left[\hat{H}_2\psi(2)\right]$$

It should be noted that the total wavefunction is the *product* of the one-electron wavefunctions, whereas the total energy is the *sum* of the one-electron energies.

The one-electron wavefunctions are similar to the hydrogen $1s$ orbitals, except that the charge on the nucleus has increased to $-Ze$, thereby making the orbitals more compact. They take the form:

$$\psi_{1s}\left(1\right) = N\,e^{-Z r_1/a_0} \quad \text{and} \quad \psi_{1s}\left(2\right) = N\,e^{-Z r_2/a_0} \qquad (7.22)$$

As before, N is the normalization constant.

The energies of electrons in these hydrogen-like orbitals are equal to $Z^2 E_{\mathrm{H}}$, where E_{H} is the energy of the hydrogen $1s$ electron ($= -13.6$ eV). The factor Z^2 is required to allow for the increased attraction of the electron to the nucleus. For the helium atom, $E_1 + E_2 = 2 \times 4 \times (-13.6)$ eV $= -108.8$ eV, which is much more negative than the experimentally determined energy of -79.0 eV.

Although we have ignored the repulsion between the two electrons in formulating the wave equation, it is possible to include the repulsion at a later stage. This is done by calculating the repulsion energy that would arise between the supposedly static charge distributions arising from two electrons in hydrogen-like $1s$ orbitals. This calculation gives a repulsion energy of 34.0 eV. When this is added to $E_1 + E_2$, we arrive at an overall energy of -74.8 eV, which is in reasonably good agreement with the experimental value. Still closer agreement can be obtained by allowing for the "screening" effect of the other electron, which tends to reduce the effective charge on the nucleus "seen" by the first electron. This is done by replacing Z in equation (7.22) with Z_{eff}, where $Z_{\mathrm{eff}} < Z$. Screening will be discussed in more detail in Section 7.7.

7.5   Electron Spin

7.5.1 The Pauli Exclusion Principle

The wavefunction that we have just derived for the helium atom is incomplete because it does not include the spins of the two electrons. The occupation of atomic oritals in many-electron atoms is controlled by the Pauli exclusion principle, which states that:

Two electrons cannot have the same set of quantum numbers.

Spin-up and spin-down refer to electron spins oriented in opposite directions. The sign given to a particular spin is arbitrary.

This set includes not only the three orbital quantum numbers, n, l and m_l, but also the spin quantum number m_s, which was discussed in Chapter 5. The spin quantum number is restricted to one of two values: $+\frac{1}{2}$, corresponding to spin-up, and $-\frac{1}{2}$, corresponding to spin-down. This leads to an alternative formulation of the Pauli exclusion principle as:

A maximum of two electrons can occupy the same spatial orbital, and then only if their spins are paired.

As we shall see, the exclusion principle is an essential part of our understanding of the structure of many-electron atoms.

The Pauli exclusion principle can be understood at a deeper level by considering what happens to the overall wavefunction when two electrons are interchanged. We can indicate this interchange by writing:

$$\Psi(1,2) \rightarrow \Psi(2,1) \qquad (7.23)$$

This process cannot affect the physical properties of the system because the electrons are indistinguishable; therefore the probability distribution, and hence $\Psi\Psi^*$, must remain unchanged. For this to be true, either the wavefunction must be unaffected by the interchange of electrons, or it must merely change sign. That is:

$$\text{either } \Psi(1,2) = \Psi(2,1) \text{ or } \Psi(1,2) = -\Psi(2,1) \qquad (7.24)$$

In practice, it is found that the electronic properties of atomic and molecular systems can be understood only if the second alternative is true. This leads to a more fundamental statement of the Pauli exclusion principle as:

When any two electrons in a many-electron system are interchanged, the total wavefunction, including the spin contribution, must change sign.

Wavefunctions which obey this principle are said to be antisymmetric.

7.5.2 Inclusion of Spin in the Wavefunctions for the Helium Atom

To see how this principle operates we need to include spin in the helium atom wavefunction by formally writing "α" for spin-up and "β" for spin-down, and indicating the electron with a particular spin by a number in parentheses. At first sight, there appear to be four spin possibilities for the two electrons in the helium atom:

$$\alpha(1)\alpha(2) \qquad \beta(1)\beta(2) \qquad \alpha(1)\beta(2) \qquad \alpha(2)\beta(1)$$

However, there is a problem with the last two assignments because they imply that it is possible to know with certainty that electron 1 has spin α and electron 2 spin β, or *vice versa*. It has already been pointed out that electrons are indistinguishable and, because of the operation of the uncertainty principle, it is not possible to trace the trajectory of a particular electron in the atom. Thus, where the electrons have opposite spins, there must always be equal probabilities of finding the configurations $\alpha(1)\beta(2)$ and $\alpha(2)\beta(1)$. This is ensured by making linear combinations of these two spin functions:

Particles that have antisymmetric wavefunctions are known as **fermions**. They include protons and ^{13}C nuclei, as well as electrons. Other particles, such as photons and ^{16}O nuclei, have symmetric wavefunctions that do not change sign when identical particles are interchanged. These are known as **bosons**.

$$\frac{1}{\sqrt{2}}\left[\alpha(1)\beta(2)+\alpha(2)\beta(1)\right] \text{ and } \frac{1}{\sqrt{2}}\left[\alpha(1)\beta(2)-\alpha(2)\beta(1)\right] \quad (7.25)$$

Here, the factor of $1/\sqrt{2}$ is required to normalize the spin functions. The four acceptable spin functions are therefore:

$$\alpha(1)\alpha(2) \quad (7.26)$$

$$\beta(1)\beta(2) \quad (7.27)$$

$$\frac{1}{\sqrt{2}}\left[\alpha(1)\beta(2)+\alpha(2)\beta(1)\right] \quad (7.28)$$

$$\frac{1}{\sqrt{2}}\left[\alpha(1)\beta(2)-\alpha(2)\beta(1)\right] \quad (7.29)$$

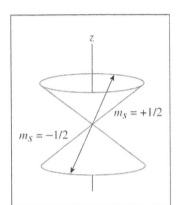

Figure 7.2 Orientation of the electron spins in the ground state of helium. The spin vectors make a definite angle with the z axis, but their position on the surface of the cones is arbitrary provided they cancel one another out

A **determinant** is a square, two-dimensional array of terms, which are cross-multiplied according to certain rules. For a 2×2 determinant the rules give:

$$\begin{vmatrix} a(1) & b(1) \\ a(2) & b(2) \end{vmatrix} = a(1)b(2) - a(2)b(1)$$

It should be noted that the first three spin functions remain the same when electrons 1 and 2 are interchanged, and they are said to be symmetric with respect to interchange of electrons. However, the fourth spin function changes sign when this is done, showing that it is antisymmetric.

In the ground state of the helium atom the spatial part of the wavefunction, $\psi_{1s}(1)\psi_{1s}(2)$, is symmetric with respect to interchange of electrons, and therefore the total wavefunction will be antisymmetric only if the spin part of the wavefunction is also antisymmetric. Thus, the overall wavefunction must be:

$$\Psi(1,2)=\psi_{1s}(1)\psi_{1s}(2)\frac{1}{\sqrt{2}}\left[\alpha(1)\beta(2)-\alpha(2)\beta(1)\right] \quad (7.30)$$

This requires the spins to be paired, as in the original statement of the Pauli exclusion principle. In this state the spins of the two electrons are oriented on their respective cones of uncertainty so that the resultant spin is zero, as shown in Figure 7.2.

Equation (7.30) can also be written in the form of a determinant:

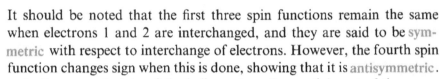

$$\Psi(1,2)=\frac{1}{\sqrt{2}}\begin{vmatrix} \psi_{1s}(1)\alpha(1) & \psi_{1s}(1)\beta(1) \\ \psi_{1s}(2)\alpha(2) & \psi_{1s}(2)\beta(2) \end{vmatrix} \quad (7.31)$$

This way of writing the wavefunction was developed by Slater in 1929, and is therefore known as a Slater determinant. Each term in the determinant consists of a hydrogen-like orbital multiplied by a spin function, and is referred to as a spin orbital. The first row in the determinant contains the two spin orbitals available to an electron in the ground state of helium, both occupied by electron 1. The second row contains the same terms, this time occupied by electron 2. The method can easily be extended to atoms containing more than two electrons by finding the appropriate determinant, but the total wavefunction cannot be written as the product of a spatial function and a spin function when more than two electrons are involved.

7.5.3 Excited States of the Helium Atom

If one of the electrons is promoted to a 2s orbital, we might expect the spatial wavefunction to be $\psi_{1s}(1)\,\psi_{2s}(2)$, but this would conflict with the indistinguishability of electrons because it implies that the orbital occupied by each electron is known. There are two ways of writing the spatial wavefunction which avoid this problem:

$$\psi_s = \frac{1}{\sqrt{2}}\left[\psi_{1s}(1)\psi_{2s}(2) + \psi_{1s}(2)\psi_{2s}(1)\right] \tag{7.32}$$

and

$$\psi_a = \frac{1}{\sqrt{2}}\left[\psi_{1s}(1)\psi_{2s}(2) - \psi_{1s}(2)\psi_{2s}(1)\right] \tag{7.33}$$

The first function is symmetric with respect to interchange of electrons and the second function is antisymmetric. When the wavefunctions are evaluated it is found that ψ_a tends to keep the two electrons apart, whereas ψ_s allows the electrons to come much closer together. It follows that ψ_a must have a lower energy than ψ_s because of the reduced inter-electron repulsion.

One simple way of demonstrating this effect is to see what happens to the wavefunctions when both electrons are equidistant from the nucleus. In this situation, $r_1 = r_2$, and $\psi_{1s}(1) = \psi_{1s}(2)$, $\psi_{2s}(1) = \psi_{2s}(2)$. The antisymmetric wavefunction, ψ_a, then becomes zero, showing that the electrons can never be at the same distance from the nucleus at the same time. This does not apply to the symmetric wavefunction, ψ_s, where there is a finite probability of finding both electrons at the same distance from the nucleus.

To satisfy the Pauli principle the antisymmetric wavefunction must be combined with one of the three symmetric spin states, given by equations (7.26)–(7.28). This particular excited state can therefore exist in three different forms. These have slightly different energies because of the small magnetic interactions which occur between the spin and orbital motions of the electrons, and this causes any spectral lines involving this state to be split into three. For this reason it is known as a triplet state. The symmetric wavefunction, ψ_s, combines with the single antisymmetric spin state, and it is said to form a singlet state.

Although it is clear that the spin states $\alpha(1)\alpha(2)$ and $\beta(1)\beta(2)$ have aligned spins, it might be thought that the function $\frac{1}{\sqrt{2}}[\alpha(1)\beta(2) + \alpha(2)\beta(1)]$ would have a resultant spin of zero because it is a combination of opposite spin states. In fact, as shown in Figure 7.3, the spin vectors are posi-

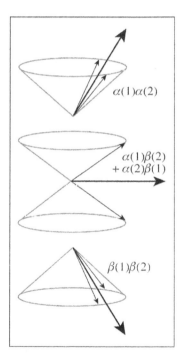

Figure 7.3 The three spin states for an excited helium atom. The total spin vector has a constant length but can adopt three different orientations with respect to the z axis

tioned in such a way that there is still a resultant spin, which is equal in magnitude to that for the other two symmetric spin states. Thus, the three spin states differ only in the orientation of the vector representing the total spin angular momentum.

7.6   The Orbital Approximation for Lithium

For an atom with n electrons, the total wavefunction can be written approximately as the product of n one-electron wavefunctions:

$$\Psi = \psi_1 \psi_2 \dots \psi_n \qquad (7.34)$$

These one-electron wavefunctions are similar to the hydrogen orbitals, but are contracted towards the nucleus because of the greater nuclear charge "seen" by the electrons. For lithium, which has three electrons, one form of the wavefunction will be:

$$\Psi(1,2,3) = \psi_{1s}(1)\alpha(1)\psi_{1s}(2)\beta(2)\psi_{2s}(3)\alpha(3) \qquad (7.35)$$

Here, electrons 1 and 2 occupy the 1s orbital with their spins paired. The third electron cannot enter this orbital because of the operation of the Pauli exclusion principle, and it goes into the 2s orbital. Unlike the situation with hydrogen, the 2s orbital has a lower energy than the 2p orbital because of electron–electron interaction (see Section 7.7), and it is occupied preferentially. The electronic structure of lithium can therefore be written as $1s^2 2s^1$. The electrons in the 1s orbital are, on average, much closer to the nucleus than the electron in the 2s orbital, and they are said to occupy the K shell.

The wavefunction given in equation (7.35) is not completely satisfactory because it assigns each electron to a particular spin orbital, whereas it is not possible to know with certainty what state an electron is in. To give all combinations equal weighting, and to make the overall wavefunction antisymmetric with respect to interchange of electrons, equation (7.35) must be rewritten as a Slater determinant:

The determinant:

$$\begin{vmatrix} a(1) & b(1) & c(1) \\ a(2) & b(2) & c(2) \\ a(3) & b(3) & c(3) \end{vmatrix}$$

is equal to:

$a(1)b(2)c(3) - a(1)b(3)c(2)$

$-a(2)b(1)c(3) + a(3)b(1)c(2)$

$+a(2)b(3)c(1) - a(3)b(2)c(1)$

$$\Psi(1,2,3) = \frac{1}{\sqrt{6}} \begin{vmatrix} \psi_{1s}(1)\alpha(1) & \psi_{1s}(1)\beta(1) & \psi_{2s}(1)\alpha(1) \\ \psi_{1s}(2)\alpha(2) & \psi_{1s}(2)\beta(2) & \psi_{2s}(2)\alpha(2) \\ \psi_{1s}(3)\alpha(3) & \psi_{1s}(3)\beta(3) & \psi_{2s}(3)\alpha(3) \end{vmatrix} \qquad (7.36)$$

This determinant gives equal weighting to the six ways in which the three electrons can be fitted into the spin orbitals, whereas equation (7.35) includes only one arrangement.

A determinant changes sign when any two rows or columns are interchanged, and the determinant in equation (7.36) is therefore automatically antisymmetric with respect to interchange of any two electrons. It is also worth noting that a determinant is equal to zero if any two rows or columns are identical. Thus, the wavefunction in (7.36) disappears if two electrons have the same set of quantum numbers.

7.7 Electron Shielding of the Nuclear Charge in Many-electron Atoms

For the hydrogen atom we found that the electron energy depended only upon the principal quantum number n, but this is no longer true for atoms containing more than one electron. For a fixed value of the quantum number n, the energies of the orbitals in many-electron atoms are generally found to be in the order s < p < d. This difference is caused by the coulombic repulsion that exists between the electrons. For a particular electron, the presence of the other electrons can be represented by a spherically symmetrical charge distribution, as shown in Figure 7.4. If the electron under consideration is at a distance r from the nucleus, it can be shown from elementary electrostatics that the repulsive effect of the other electrons is equivalent to a negative point charge, located at the nucleus and equal in magnitude to the total electronic charge contained within a sphere of radius r. The electronic charge outside this sphere makes no net contribution to the repulsive force. This inter-electron repulsion has to be averaged over all possible values of r, and the overall effect can be represented by an imaginary negative charge at the nucleus equal to σe, where σ is known as the screening constant. Thus, the effective number of positive charges that the electron "sees" at the nucleus is reduced from Z to $(Z - \sigma)$.

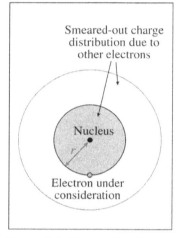

Figure 7.4 Diagram showing the effect of electron–electron repulsions on the effective nuclear charge experienced by an electron situated at a distance r from the nucleus

It should be emphasized that the screening constant, σ, will not be the same for electrons in different orbitals of the same atom because of the varying extents to which electrons approach the nucleus. σ can be thought of as the average number of electronic charges that come between the particular electron under consideration and the nucleus. For electrons that never come close to the nucleus, σ will be large, but for electrons that penetrate inner electron shells it will be much smaller.

For a given value of the principal quantum number, n, electrons in s orbitals are more likely to be found very close to the nucleus than electrons in p orbitals, and therefore the effective nuclear charge that they experience is greater. For this reason, an s electron is more tightly bound than the equivalent p electron.

The radial probability distributions for the hydrogen 2s and 2p orbitals are compared in Figure 7.5. It can be seen that the maximum electron density for the 2s orbital occurs at a slightly greater distance from the nucleus than it does for the 2p orbital, but this is counterbalanced by the 2s orbital having a subsidiary peak, which is very close to the nucleus. Taken overall, the 2s orbital penetrates the inner 1s shell to a greater extent than the 2p orbital does. In many-electron atoms the orbitals will be smaller than their hydrogen equivalents because of the increased nuclear charge, but the relative penetration of one orbital by another will remain roughly the same.

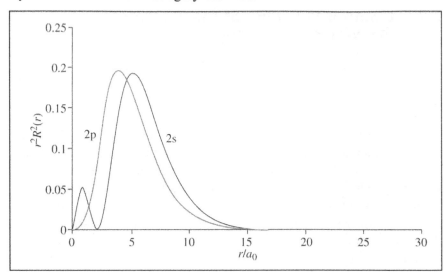

Figure 7.5 The radial probability distributions for electrons in 2s and 2p orbitals

For $n = 3$, we find that electrons in the 3d orbitals lie much further from the nucleus on average than electrons in either the 3s or 3p orbitals, with the result that the order of the energies is

$$3s < 3p < 3d$$

To a rough approximation the ionization energy, ε, of an electron in a hydrogen-like orbital can be related to the screening constant, σ, by the formula:

$$\varepsilon = 13.6 \times \left[\frac{Z - \sigma}{n} \right]^2 \text{ eV} \qquad (7.37)$$

Here, n is the principal quantum number relating to the orbital, and 13.6 eV is the ionization energy of an electron in the 1s level of the hydrogen atom.

Worked Problem 7.2

Q The ionization energy for the outer 2s electron of the lithium atom is 5.39 eV. Calculate the effective nuclear charge experienced by this electron.

A Using equation (7.37), we have:

$$\left[\frac{Z-\sigma}{2}\right]^2 = \frac{5.39 \text{ eV}}{13.6 \text{ eV}} = 0.396$$

This gives:

$$Z - \sigma = 2\sqrt{0.396} = 1.26$$

Thus, the number of positive charges that the electron "sees" at the nucleus has been reduced from 3.00 to 1.26 by the very effective screening provided by the 1s electrons.

7.8 The Use of Self-consistent Field Methods to Obtain Atomic Orbitals

In a many-electron atom an approximate wavefunction, ϕ_i, can be obtained for the ith electron by solving the following one-electron wave equation:

$$\left[-\frac{\hbar^2}{2m_e}\nabla_i^2 + V(r_i)\right]\phi_i = \varepsilon_i\phi_i \qquad (7.38)$$

The first term on the left-hand side represents the kinetic energy of the electron, and the second term the potential energy. ε_i is the energy of the electron in the ith orbital. The potential energy term, $V(r_i)$, is calculated by assuming that the electrical repulsion resulting from all the other electrons can be represented by a static, spherically symmetric charge distribution. A difficulty with this procedure is that it is impossible to calculate $V(r_i)$ until the one-electron wavefunctions of all the other electrons are known. Thus, all the one-electron wavefunctions derived from equation (7.38) are interrelated, which makes the problem difficult to solve.

In 1928, D. R. Hartree got around this difficulty by adopting the following procedure. First, all the one-electron wavefunctions were estimated, using effective nuclear charges. These wavefunctions were then used to calculate the potential energy term for the first electron. This was substituted into equation (7.38), and an improved wavefunction for electron 1 was calculated. This was then used to calculate the potential energy term for the second electron, and hence to improve its wavefunction. This process was repeated for all the electrons until self-consistency was obtained, that is, further repetition of the process did not produce any changes in the wavefunctions.

Hartree incorporated the Pauli principle by allowing no more than two electrons to be present in each orbital, but the wavefunctions that he used did not involve spin, and were not antisymmetric with respect to interchange of electrons. In 1930, V. Fock modified Hartree's approach by using fully antisymmetric spin orbitals that did not distinguish between electrons. This improved way of calculating atomic orbitals is known as the Hartree–Fock self-consistent field (SCF) method. Nowadays, fast computers are used and procedures are followed which allow the one-electron wave equations to be solved simultaneously.

7.9   Electron Correlation Energy

A weakness of the Hartree–Fock SCF method is that it does not take proper account of the coulombic repulsion that exists between electrons, and which causes them to avoid one another. This is known as electron correlation, and its neglect causes the total energy of a typical atom to be overestimated by an amount which can be of the order of 100 kJ mol^{-1}. This energy is known as the correlation energy, and it is very difficult to make proper allowance for it in calculations. Some progress has been made by using so-called configurational interactions, in which atomic orbitals for both occupied and excited states of the atom are used, but these calculations require very large amounts of computer time.

7.10 The Elements of the Periodic Table

Atomic orbital theory, together with the Pauli exclusion principle, provides an explanation for the ordering of the elements in the Periodic Table. The electron configuration of each element in its ground state can be obtained by feeding electrons into the hydrogen-like atomic orbitals in order of increasing energy until the full compliment of Z electrons per atom is reached. Each orbital can accommodate two electrons with their spins paired. This procedure is known as the Aufbau or building-up principle.

The electron configurations of the first 36 elements are given in Table 7.1. Here [He], [Ne] and [Ar] are used to indicate the closed-shell configurations of the first three inert gases, helium, neon and argon. The general order in which the orbitals are occupied is found to be

1s 2s 2p 3s 3p 4s 3d 4p 5s 4d 5p 6s 4f 5d 6p

To a large extent this follows the order of increasing energies for the one-electron orbitals, as calculated by the Hartree–Fock SCF method. However, there can be complications when two orbital energies have similar values, as happens for example with the 3d and 4s orbitals of elements between potassium and nickel. Under these conditions the order in which the orbitals are filled depends upon the particular electronic configuration of the atom because this has an effect on the orbital energies.[1,2] This topic will be discussed in more detail when the transition metal series is examined.

Table 7.1 The electron configurations of the first 36 elements

Z	Atom	Electron structure	Z	Atom	Electron structure
1	H	$1s^1$	19	K	$[Ar]4s^1$
2	He	$1s^2$	20	Ca	$[Ar]4s^2$
3	Li	$[He]2s^1$	21	Sc	$[Ar]4s^23d^1$
4	Be	$[He]2s^2$	22	Ti	$[Ar]4s^23d^2$
5	B	$[He]2s^22p^1$	23	V	$[Ar]4s^23d^3$
6	C	$[He]2s^22p^2$	24	Cr	$[Ar]4s^13d^5$
7	N	$[He]2s^22p^3$	25	Mn	$[Ar]4s^23d^5$
8	O	$[He]2s^22p^4$	26	Fe	$[Ar]4s^23d^6$
9	F	$[He]2s^22p^5$	27	Co	$[Ar]4s^23d^7$
10	Ne	$[He]2s^22p^6$	28	Ni	$[Ar]4s^23d^8$
11	Na	$[Ne]3s^1$	29	Cu	$[Ar]4s^13d^{10}$
12	Mg	$[Ne]3s^2$	30	Zn	$[Ar]4s^23d^{10}$
13	Al	$[Ne]3s^23p^1$	31	Ga	$[Ar]4s^23d^{10}4p^1$
14	Si	$[Ne]3s^23p^2$	32	Ge	$[Ar]4s^23d^{10}4p^2$
15	P	$[Ne]3s^23p^3$	33	As	$[Ar]4s^23d^{10}4p^3$
16	S	$[Ne]3s^23p^4$	34	Se	$[Ar]4s^23d^{10}4p^4$
17	Cl	$[Ne]3s^23p^5$	35	Br	$[Ar]4s^23d^{10}4p^5$
18	Ar	$[Ne]3s^23p^6$	36	Kr	$[Ar]4s^23d^{10}4p^6$

Electrons with principal quantum number, n, equal to 1, 2, 3 and 4, are said to occupy the K, L, M and N shells, respectively. The K shell is full at helium and the next electron goes into a 2s orbital, giving lithium the electron configuration $1s^22s^1$, as we have already seen. This is followed by beryllium, which has the electron configuration $1s^22s^2$. Further electrons must enter the 2p orbitals, where a total of six electrons can be accommodated. In nitrogen the three p electrons occupy separate orbitals because this keeps them apart, thereby reducing the coulombic repulsion between them, and the ground state configuration is therefore $1s^22s^22p_x^12p_y^12p_z^1$.

The L shell is full at neon, and the next electron goes into a 3s orbital, giving sodium the configuration $[Ne]3s^1$. Further electrons go into the 3s and 3p orbitals until argon is reached, which is classed as having a closed shell, even though the M shell is incomplete. Argon is unreactive because the energy required to promote an electron from the occupied 3s and 3p orbitals into the empty 3d orbitals is very great and inhibits strong bond formation.

The next element, potassium, is of interest because the extra electron goes into the 4s orbital, thereby showing that the $[Ar]4s^1$ configuration has a lower energy than the [Ar]$3d^1$ configuration. This reversal of the expected order comes about because the 4s electron penetrates further into the argon core than the 3d electron does, and therefore experiences a greater force of attraction towards the nucleus. The same effect is found in calcium, which has the configuration [Ar]$4s^2$.

Further electrons go into 3d orbitals to give the first series of transition metals which, with two exceptions, have the general configuration [Ar]$4s^2 3d^n$. Here, there is good evidence that the energy of the 4s orbital lies above that of the 3d orbital. For example, spectroscopic data show that the dipositive ions of these metals all have the general configuration [Ar]$3d^n$, from which it can be deduced that an electron is more easily removed from a 4s orbital than it is from a 3d orbital. This is in keeping with Hartree–Fock SCF calculations which show that the energy of a 3d orbital always lies below that of a 4s orbital in atoms where both orbitals are occupied.

On this basis the preferential occupation of the 4s orbital in transition metals needs some explanation, because the 3d orbitals might be expected to fill first. If scandium is taken as an example, we might expect [Ar]$3d^3$ to be the most stable configuration, but spectroscopy shows that this state lies 404 kJ mol^{-1} above the ground state, which has the configuration [Ar]$4s^2 3d^1$. The problem disappears once it is realized that the energy of the 3d orbital will be considerably greater in the [Ar]$3d^3$ configuration than it is in the ground state configuration, because of the large coulombic repulsion that exists between the three electrons in the 3d orbitals.

In situations like this, where the energies separating different orbitals are quite small, the one-electron approximation breaks down, and the electron configuration of the ground state can then be determined only by looking at the atom as a whole and treating the electrons collectively.

7.11 Hund's Rule

In Section 7.5.3 we saw that the $1s^1 2s^1$ configuration of the helium atom can exist in two forms: one where the electron spins are parallel, and the other where the spins are paired. The state with parallel spins was found to have the lower energy because the antisymmetric spatial wavefunction required the electrons to stay further apart, thereby reducing the coulombic repulsion energy.

The same principle can be applied to other multi-electron systems. In general, the coulombic repulsion between electrons is found to be least when the number of electrons with parallel spins has been maximized. These conclusions are in accordance with Hund's rule, which states that:

For an atom with a particular electron configuration, the most stable state is the one which has the maximum number of unpaired electrons.

The application of this principle to a variety of elements is shown in

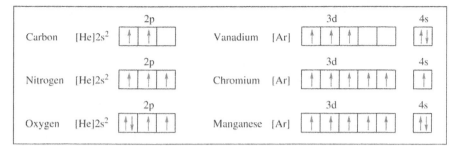

Figure 7.6. In nitrogen, which has the electron configuration $[He]2s^2 2p_x^1 2p_y^1 2p_z^1$, the spins of the three 2p electrons will all be parallel in the ground state. The next element, oxygen, has four 2p electrons, two of which must go into the same orbital with their spins paired. This leaves just two unpaired electrons with parallel spins.

The transition metals are worthy of note because the reduction in electron repulsion gained from parallel spins makes the $3d^5$ configuration particularly stable. For chromium, this is sufficient to make the transfer of an electron from the 4s orbital into one of the 3d orbitals energetically favourable. This results in an $[Ar]4s^1 3d^5$ configuration in which both the 3d and the 4s orbitals are only partially occupied, and the simple Aufbau principle breaks down.

Worked Problem 7.3

Q Find the number of unpaired electrons in the ground state of a nickel atom.

A From Table 7.1, nickel has the electron configuration $[Ar]4s^2 3d^8$. Three d orbitals will be doubly occupied by electrons with their spins paired, and the remaining two d orbitals will contain single electrons with their spins parallel. Thus, there will be two parallel spins.

7.12 Ionization Energies of the Elements

The first ionization energy of an atom, A, is defined as the energy required to remove an electron completely from the isolated, neutral atom. It can be represented by the equation:

$$A \rightarrow A^+ + e^- \tag{7.39}$$

It is a measure of the strength of the forces holding the electron to the atom. Further electrons can be removed from the ion A^+, leading to second and higher ionization energies.

The way in which the first ionization energy varies with atomic number for the first 55 elements is shown in Figure 7.7. It can be seen that the ionization energies follow a regular sequence, with the highest values occurring at the inert gases and the lowest values at the alkali metals. This behaviour can be explained by the sequential filling of the electron energy levels. The alkali metals have one electron outside a filled shell, and this electron is easily removed because the electrons in the inner shells screen the nuclear charge so effectively that the outer electron "sees" only a small positive charge attracting it towards the nucleus. Thus, as we have already seen, the effective nuclear charge experienced by the outermost 2s electron in lithium is reduced from a nominal $-3e$ to $-1.3e$ because of the screening effect of the two 1s electrons.

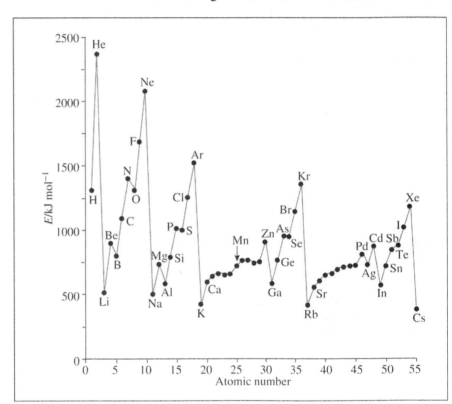

Figure 7.7 Variation of the first ionization energy of the elements with atomic number

As more electrons are introduced into the outer shell, the effective nuclear charge rises because electrons in the same shell are unable to screen the nuclear charge from one another very effectively, and the ionization energy therefore increases. This effect reaches its climax when the shell is full, which occurs at neon for the L shell.

It should be noted that the rise in first ionization energy from lithium to neon is not continuous, but shows small reversals at boron and oxygen. At boron the 2p orbitals start to be occupied, and a small fall in ionization energy occurs because electrons in 2p orbitals are less strongly bound than electrons in 2s orbitals. A different explanation is needed for the small fall in ionization energy observed at oxygen. Nitrogen comes before oxygen, and it has three 2p electrons which occupy separate $2p_x$, $2p_y$ and $2p_z$ orbitals. The addition of a fourth electron at oxygen causes one of the 2p orbitals to become doubly occupied, and the presence of two electrons in the same orbital gives rise to considerable coulombic repulsion, thereby lowering the ionization energy.

It can be seen that the ionization energy of the alkali metals decreases as we go from lithium to cesium. This occurs because the electron in the outermost s orbital is getting further away from the nucleus.

Summary of Key Points

1. The spatial wavefunction, $\Psi(1,2)$, for the helium atom can be written approximately as the product of two hydrogen-like wavefunctions:

$$\Psi\big(1,2\big) = \psi_{1s}\big(1\big)\psi_{1s}\big(2\big)$$

This is known as the one-electron approximation. The numbers in parentheses refer to the two electrons.

2. The Pauli Exclusion Principle states that the wavefunction for a system of electrons must be antisymmetric with respect to interchange of any two electrons. This requires the two electrons in the ground state of helium to have opposite spins. It also explains why the first excited state of helium exists in three forms, known as a triplet state.

3. For hydrogen, orbitals with the same principal quantum number have the same energy, but this is no longer true for many-electron atoms because of the effect of electron–electron repulsions which cause the effective nuclear charge "seen" by a particular electron to be different for electrons in different orbitals. Thus, for a given value of the principal quantum number n, the orbital energies tend to be in the order: s < p < d.

4. The energies of the one-electron wavefunctions can be calculated by the Hartree–Fock method in which the repulsive effect of the electrons on one another is calculated by a process of successive approximation.

5. The ground state electron configuration of each element in the Periodic Table can be obtained by feeding two electrons, with spins paired, into each of the one-electron orbitals in order of increasing energy until the full complement of electrons per atom is reached. This is known as the Aufbau Principle. It can break down when there are two types of orbital with similar energies, and the correct electron configuration of the ground state can then be determined only by looking at the atom as a whole and treating the electrons collectively.

6.   Where there are states which differ only in the number of unpaired electrons, Hund's Rule states that the state with the maximum number of parallel electron spins has the lowest energy.

Problems

7.1. Write down the electronic Hamiltonian operator for a system consisting of two non-interacting hydrogen atoms, A and B, placed a large distance apart. The answer should be given in terms of the distance of electron 1 from nucleus A, r_{1A}, and the distance of electron 2 from nucleus B, r_{2B}. What will the wavefunction of such a system be?

7.2. Write down the spin-orbital wavefunction for the singlet state of an excited helium atom in which one electron is in a 1s orbital and the other is in a 2s orbital.

7.3. By multiplying out the terms, show that the following wavefunction is antisymmetric with respect to interchange of electrons 1 and 2:

$$\Psi = \begin{vmatrix} \psi_a(1) & \psi_a(2) \\ \psi_b(1) & \psi_b(2) \end{vmatrix}$$

7.4. The first ionization energies of sodium, potassium and rubidium atoms are 5.14, 4.34 and 4.17 eV, respectively. Use equation (7.37) to calculate the effective nuclear charge experienced by the outermost s electron in each of these atoms.

7.5. The second ionization energy of lithium is 75.6 eV. This corresponds to the process:

$$Li^+ \rightarrow Li^{2+} + e^-$$

Calculate the effective nuclear charge experienced by an electron in the 1s orbital and comment on the result.

7.6. The single-bond covalent radii of Na and Mg are 0.154 and 0.136 nm, respectively. Thus, the magnesium atom is slightly smaller than the sodium atom. It could be argued that adding a second 3s electron would increase the size of the atom because of the electrical repulsion existing between the two outer electrons. Explain why a contraction is observed instead.

7.7. Hartree–Fock self-consistent field calculations indicate that the energy of an electron in the 4s orbital of vanadium lies above that of the 3d orbital in the ground state configuration, $[Ar]3d^34s^2$. Explain why $[Ar]3d^44s^1$ and $[Ar]3d^5$ are less stable configurations than the ground state.

7.8. Use the Pauli exclusion principle and Hund's rule to find the number of unpaired electrons in atoms of the following elements: fluorine ($Z = 9$), phosphorus ($Z = 15$), titanium ($Z = 22$) and cobalt ($Z = 27$).

7.9. Explain why the first ionization energy of sulfur ($Z = 16$) is less than that of phosphorus ($Z = 15$), despite a generally rising trend with atomic number from $Z = 11$ to $Z = 18$.

References

1. M. P. Melrose and E. R. Scerri, *J. Chem. Educ.*, 1996, **73**, 498.
2. L. G. Vanquickenborne, K. Pierloot and D. Devoghel, *J. Chem. Educ.*, 1994, **71**, 469.

Further Reading

D. A. McQuarrie and J. D. Simon, *Physical Chemistry: A Molecular Approach*, University Science Books, Sausalito, California, 1997, chap. 8.
P. W. Atkins and R. S. Friedman, *Molecular Quantum Mechanics*, 3rd edn., Oxford University Press, Oxford, 1997, chap. 7.

8

The Structure of Molecules

By the end of this chapter you should be able to:

- Understand how the electronic and nuclear wavefunctions of a molecule can be separated, using the Born–Oppenheimer approximation
- Compare and contrast the molecular orbital and valence bond descriptions of the bonding in the hydrogen molecule
- Understand the importance of electron correlation in formulating effective wavefunctions
- Estimate the energies of the bonding and antibonding molecular orbitals of diatomic molecules from the secular determinant
- Construct simple molecular orbitals for diatomic molecules from a linear combination of atomic orbitals and describe their symmetry
- Construct molecular orbitals for heteronuclear diatomics such as NO and HF
- Explain the bonding in molecules such as BeH_2, BH_3, CH_4 and NH_3 in terms of the overlap of hydrogen 1s orbitals with hybrid atomic orbitals formed on the central atom
- Compare valence bond and molecular orbital descriptions of the bonding in the water molecule
- Use Hückel molecular orbital theory to construct molecular orbitals for conjugated hydrocarbon molecules such as butadiene and benzene

8.1   Introduction

The first quantitative theory of chemical bonding was developed for the hydrogen molecule by Heitler and London in 1927, and was based on the Lewis theory of valence in which two atoms shared electrons in such a way that each achieved a noble gas structure. The theory was later extended to other, more complex molecules, and became known as valence bond theory . In this approach, the overlap of atomic orbitals on neighbouring atoms is considered to lead to the formation of localized bonds, each of which can accommodate two electrons with paired spins. The theory has been responsible for introducing such important concepts as hybridization and resonance into the theory of the chemical bond, but applications of the theory have been limited by difficulties in generating computer programs that can deal efficiently with anything other than the simplest of molecules.

An alternative approach to bonding uses the concept of one-electron orbitals, which was developed for atoms in Chapter 7. This is known as molecular orbital (MO) theory. The one-electron wavefunctions are considered to be spread over the whole molecule, leading to the formation of molecular orbitals. Unlike the valence bond approach, this theory can be applied quite generally to a wide variety of molecules without the need to introduce specific features for individual molecules. Once the relative energies of the molecular orbitals have been calculated, electrons are fed into the lowest lying orbitals to give the overall electronic structure of the molecule.

In this chapter we will focus on MO theory because this is the most widely used method of calculating molecular properties, but valence bond theory will be discussed where appropriate. Before entering into a detailed discussion of the molecular orbitals of simple diatomic molecules, it will be useful to delve a little deeper into quantum mechanics, and take a look at ways of evaluating approximate solutions of the Schrödinger equation.

8.2   Trial Wavefunctions and their Associated Energies

In Chapter 7 we saw that it is not possible to obtain exact solutions of the Schrödinger equation for many-electron atoms, even within the one-electron approximation, and the same applies to molecules. For these systems it is necessary to use approximate solutions, which are based on chemical insight and chosen for mathematical convenience.

First, we consider an atom or molecule which has a set of wavefunctions, ψ_n, given by the equation:

$$\hat{H}\psi_n = E_n\psi_n \tag{8.1}$$

For a given Hamiltonian operator there will be an infinite number of solutions to this equation, each indicated by a different value of the index n. We wish to find the ground state wavefunction, ψ_0, which has an energy E_0. Normally, equation (8.1) cannot be solved analytically and the wavefunctions that satisfy the equation are unknown. Under these circumstances it is necessary to formulate a trial wavefunction, ϕ, which is expected to be a good approximation to the true ground state wavefunction.

One problem with this trial wavefunction is that it will not be an eigenfunction of the Hamiltonian operator for the atom or molecule. Thus:

$$\hat{H}\phi \neq \left(\text{constant}\right)\phi \tag{8.2}$$

It is evident that the energy associated with the trial wavefunction is not a constant of the motion. Any attempt to measure the energy of such a hypothetical system would force it into one of the quantum states represented by equation (8.1), and the energy measured could be any of the values E_0, E_1, ... E_n. It follows that there must be an uncertainty in the energy of such a system, and we have to resort to an average energy, $\langle E \rangle$, which is defined as the average value that would be obtained from a large number of measurements. This is often referred to as the expectation value of the energy, and it can be obtained from the equation:

$$\langle E \rangle = \frac{\int \phi^* \hat{H}\phi \, \mathrm{d}\tau}{\int \phi^* \phi \, \mathrm{d}\tau} \tag{8.3}$$

Here, $\mathrm{d}\tau$ is a shorthand notation for $\mathrm{d}x\mathrm{d}y\mathrm{d}z$, and integration is carried out over all space.

This important equation is based on one of the basic postulates of quantum mechanics, and it cannot be derived from a more fundamental equation. However, we can carry out a check on the equation by replacing the trial wavefunction with one of the eigenfunctions of the Hamiltonian operator, ψ_n. Application of equation (8.1) then leads to the correct result:

$$\langle E \rangle = \frac{\int \psi_n^* \hat{H}\psi_n \, \mathrm{d}\tau}{\int \psi_n^* \psi_n \, \mathrm{d}\tau} = \frac{E_n \int \psi_n^* \psi_n \, \mathrm{d}\tau}{\int \psi_n^* \psi_n \, \mathrm{d}\tau} = E_n \tag{8.4}$$

8.3 The Variation Principle

This can be stated as follows:

For any trial wavefunction, the expectation value of the energy can never be less than the true ground state energy.

This can be expressed mathematically as follows:

$$E_0 \leq \frac{\int \phi^* \hat{H} \phi \, d\tau}{\int \phi^* \phi \, d\tau} \qquad (8.5)$$

where E_0 is the true ground state energy, obtained from equation (8.1). The equality applies only when ϕ is identical to the true ground state wavefunction, ψ_0.

This result comes about because, as mentioned in Section 8.2, the expectation value of the energy is a weighted mean of the true energies of the system, E_0, E_1, E_2, ..., and this mean value cannot be less than E_0.

The variation principle allows different trial wavefunctions to be evaluated on the basis that the one giving the lowest energy will be the best approximation to the true ground state wavefunction. Some caution is necessary when applying this principle because trial wavefunctions which deviate significantly from the true wavefunction can sometimes give energies that are surprisingly close to the true ground state energy.

8.4 The Hamiltonian Operator for the Hydrogen Molecule-ion

We begin our consideration of chemical bonding by looking at the simplest possible molecule, H_2^+. The molecular orbitals derived for this system form the basis of the molecular orbitals for all other diatomic molecules, in much the same way that the atomic orbitals of hydrogen form the basis for all atomic orbitals.

The hydrogen molecule-ion contains two protons of mass m_p, which we shall label A and B, and a single electron of mass m_e. The distances separating the three particles are defined in Figure 8.1. The Hamiltonian operator for this system can be written as:

$$\hat{H} = -\frac{\hbar^2}{2m_p}\left(\nabla_A^2 + \nabla_B^2\right) - \frac{\hbar^2}{2m_e}\nabla_e^2 - \frac{e^2}{4\pi\varepsilon_0}\left(\frac{1}{r_A} + \frac{1}{r_B}\right) + \frac{e^2}{4\pi\varepsilon_0 R} \qquad (8.6)$$

In this equation ∇_A^2, ∇_B^2 and ∇_e^2 are Laplacian operators involving the coordinates of nucleus A, nucleus B and the electron, respectively. The first term on the right-hand side of the equation gives the kinetic energy of the two protons, and the second term gives the kinetic energy of the electron. The third term represents the electrostatic interaction energy existing between the electron and the two nuclei, and the fourth term represents the internuclear repulsion energy.

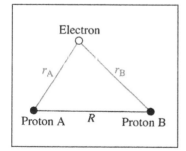

Figure 8.1 Schematic diagram of the hydrogen molecule-ion. The distances of the electron from the two nuclei are r_A and r_B, and the nuclei are separated by a fixed distance R

The Schrödinger equation with this particular Hamiltonian operator contains nine independent variables, and it is therefore much too complicated for mathematical analysis, but the complexity of the problem can be greatly reduced by separating the motions of the protons and the electron. The way in which this is done is discussed in the next section.

8.5   The Born–Oppenheimer Approximation

Electrons are much lighter than nuclei, and therefore move more quickly. For example, calculations show that the average speed of an electron in the hydrogen molecule-ion is approximately 1000 times the average speed of a proton. This means that an electron can make a complete orbit of the molecule before the nuclei have moved significantly. This enables electrons to adjust their orbitals almost instantaneously in response to any change in the positions of the two nuclei, and the motion of the nuclei (representing translation, vibration and rotation of the molecule) can therefore be separated from the electronic motion. This is known as the Born–Oppenheimer approximation, and it allows the Schrödinger equation for electronic motion to be solved for a fixed separation of the protons.

With this approximation the total wavefunction for the molecule-ion can be written as the product of an electronic wavefunction, ψ_e, which is a function only of the electron coordinates, and a nuclear wavefunction, ψ_n, which is a function only of the nuclear coordinates:

$$\Psi = \psi_e \psi_n \qquad (8.7)$$

The Schrödinger equation for the electronic motion can be written as:

$$\hat{H}_e \psi_e = E_e \psi_e \qquad (8.8)$$

where

$$\hat{H}_e = -\frac{\hbar^2}{2m_e}\nabla_e^2 - \frac{e^2}{4\pi\varepsilon_0}\left(\frac{1}{r_A} + \frac{1}{r_B}\right) + \frac{e^2}{4\pi\varepsilon_0 R} \qquad (8.9)$$

The only variables in this Hamiltonian are the coordinates of the electron, which appear in r_A, r_B and ∇_e^2. Although the internuclear separation, R, is treated as a constant when the electronic Schrödinger equation is being solved, the wavefunction obtained will depend upon the value of R used.

8.6 Molecular Orbitals for the Hydrogen Molecule-ion

Although it is possible to obtain exact solutions to the simplified Schrödinger equation given in equations (8.8) and (8.9), the resulting wavefunction is complicated, and it provides little insight into the wavefunctions that might be used for other diatomic molecules. For this reason it will be more instructive to examine trial wavefunctions that have been constructed by a linear combination of hydrogen 1s atomic orbitals:

$$\phi_{+} = N_{\pm}\left(\psi_A \pm \psi_B\right) \tag{8.10}$$

Here, ψ_A and ψ_B are hydrogen 1s orbitals centred on atoms A and B, respectively, and $N_{\pm}$ is a normalization constant, which will have different values for ϕ_{+} and ϕ_{-}. This way of constructing molecular orbitals is known as the linear combination of atomic orbitals, or LCAO method for short.

The use of this type of trial wavefunction can be justified by the following argument. When the electron is very close to nucleus A, it experiences a coulombic attraction towards nucleus A which is far greater than that towards nucleus B. The wavefunction in this region is therefore expected to resemble a hydrogen 1s orbital, centred on nucleus A. Similarly, when the electron is very close to nucleus B, the wavefunction is expected to resemble a 1s orbital centred on nucleus B. Thus, by combining these two atomic wavefunctions it should be possible to produce trial wavefunctions which are fairly close to the true wavefunctions.

The normalized atomic orbitals to be used in equation (8.10) are:

$$\psi_A = \frac{1}{\sqrt{\pi a_0^3}}e^{-r_A/a_0} \quad \text{and} \quad \psi_B = \frac{1}{\sqrt{\pi a_0^3}}e^{-r_B/a_0} \tag{8.11}$$

Combination of these two atomic orbitals results in the wavefunctions shown in Figure 8.2. Addition of the atomic orbitals leads to constructive interference in the region between the nuclei, and the wavefunction is reinforced there. On the other hand, subtraction of the orbitals produces a node between the nuclei, showing that the electrons avoid this region.

The expectation values for the energies of these trial wavefunctions can be calculated by inserting the wavefunctions into equation (8.3). This gives:

$$\langle E \rangle = N_{\pm}^2 \int \left(\psi_A \pm \psi_B\right)\hat{H}_e\left(\psi_A \pm \psi_B\right)d\tau \tag{8.12}$$

By evaluating the integrals in this equation for a series of different, but

Although ψ_A and ψ_B are both 1s atomic wavefunctions, they are functions of different variables. For ψ_A the variable is r_A, the distance of the electron from nucleus A, and for ψ_B it is r_B, the distance of the electron from nucleus B.

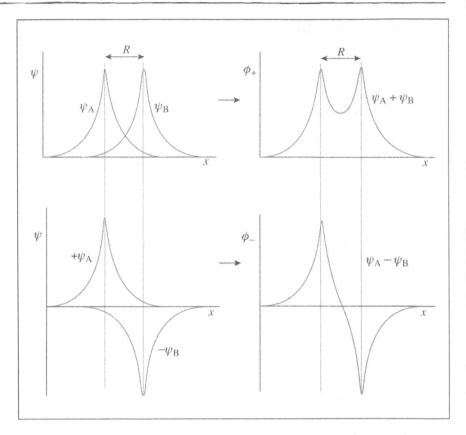

Figure 8.2 The bonding and antibonding wavefunctions that result from the overlap of two hydrogen 1s atomic wavefunctions. The distance x is measured along the internuclear axis

fixed, values of the internuclear separation R, it is possible to calculate how the energy of the molecule-ion varies with bond length. Plots of energy versus R, known as potential energy curves, are shown in Figure 8.3. As the two atoms are brought closer together, the potential energy of ϕ_+ drops to a minimum value, before rising at small values of R as the overall interaction becomes repulsive. This wavefunction must therefore represent a bonding orbital. By contrast, the potential energy of ϕ_- rises continuously as the internuclear separation is reduced, and it must therefore represent an antibonding orbital.

The depth of the potential well formed with the bonding orbital gives a measure of the bond dissociation energy of the molecule-ion. This is calculated to be +170 kJ mol^{-1}, which is much less than the true value of +258 kJ mol^{-1}. Thus, the simple molecular orbital given in equation (8.10) does not provide an accurate value for the bond dissociation energy. The calculation can be improved by combining a mixture of 1s, 2s and 2p atomic orbitals, centred on the two nuclei. In the limit, as more and more terms are added, it is possible to obtain an energy which is exactly equal to the true bond dissociation energy.

The probability density for the electron in one of the molecular orbitals can be obtained by squaring the wavefunction:

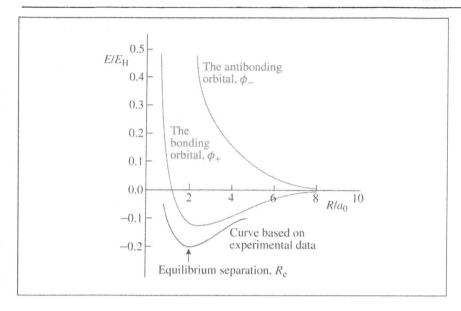

Figure 8.3 Potential energy curves for the bonding and antibonding orbitals of H_2^+ in the LCAO approximation. The true bonding curve is given for comparison. Energies are expressed as a fraction of E_H, the energy of the hydrogen 1s atomic orbital

$$\phi_{\pm}^2 = N_{\pm}^2 \left(\psi_A \pm \psi_B \right)^2$$
$$= N_{\pm}^2 \left(\psi_A^2 + \psi_B^2 \pm 2\psi_A \psi_B \right) \tag{8.13}$$

The resulting probability densities are shown in Figure 8.4. The first two terms in equation (8.13) correspond to the electron density of separate atomic orbitals centred on nuclei A and B. The third term is a measure of the excess, or deficit, electron density resulting from the formation of the combined orbital. It will be important only in the region between the nuclei, where ψ_A and ψ_B both have significant values. This term is positive for the bonding orbital, showing that there is an accumulation of

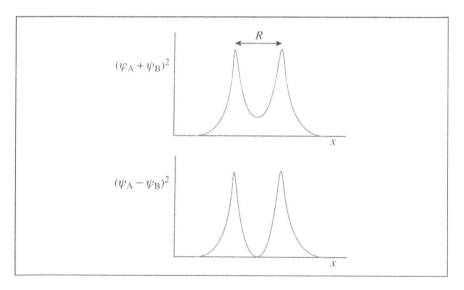

Figure 8.4 Plots of ψ^2 versus distance x along the internuclear axis for the bonding and antibonding orbitals of H_2^+

electron density in the region between the two nuclei. This is generally thought to be responsible for bonding because the electron in this region is attracted towards both nuclei, which lowers its energy. Regrettably, this simple explanation of bonding is not entirely correct, and the reader is referred to Atkins[1] for a more subtle explanation. For the antibonding orbital the third term is negative, confirming that the electrons tend to avoid the region between the two nuclei.

The bonding and antibonding molecular orbitals both have cylindrical symmetry about the internuclear axis, as shown in Figure 8.5. Orbitals of this type are known as σ orbitals. The bonding orbital also has a centre of symmetry (see Figure 8.6), which means that the magnitude of the wavefunction remains the same when the wavefunction is inverted through the centre of the molecule. Wavefunctions with this property are designated by the subscript g for *gerade* (German for even). Thus, the bonding orbital is referred to as a $1s\sigma_g$ orbital. The magnitude of the antibonding wavefunction changes sign on inversion through the centre of the molecule, and it is given the subscript u for *ungerade*. It is also given an asterisk to indicate that it is an antibonding orbital, making it a $1s\sigma_u^*$ orbital .

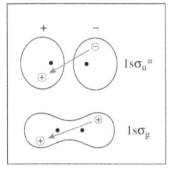

Figure 8.5 Contour plots of the wavefunctions corresponding to bonding and antibonding orbitals of H_2^+

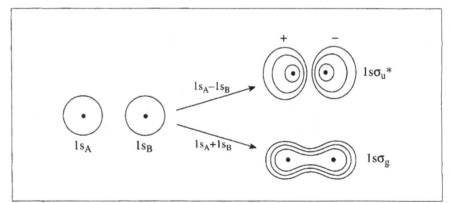

Figure 8.6 The effect on wavefunctions of inverting the electron coordinates through the centre of the molecule. The bonding wavefunction is unchanged by this process but the antibonding wavefunction changes sign

The atomic 1s wavefunctions are real, and therefore $\phi_\pm^* = \phi_\pm$.

Worked Problem 8.1

Q Derive expressions for the normalization constants of the trial wavefunctions, ϕ_+ and ϕ_-, in equation (8.10).

A We must find the values of $N_\pm$ which make:

$$\int \phi_\pm^* \phi_\pm d\tau = \int \phi_\pm^2 d\tau = 1$$

After substituting for $\phi_\pm$ from equation (8.10), we obtain:

After substituting for $\phi_{\pm}$ from equation (8.10), we obtain:

$$N_{\pm}^2\left\{\int\psi_A^2\,d\tau+\int\psi_B^2\,d\tau\pm2\int\psi_A\,\psi_B\,d\tau\right\}=1 \qquad (8.14)$$

Since the atomic orbitals are normalized:

$$\int\psi_A^2\,d\tau=1 \text{ and } \int\psi_B^2\,d\tau=1$$

The third integral in equation (8.14) is known as the overlap integral because it measures the extent to which the two atomic wavefunctions overlap one another (see Figure 8.7). It is given the symbol S_{AB}. Hence:

$$S_{AB}=\int\psi_A\,\psi_B\,d\tau \qquad (8.15)$$

The normalization constants are then found to be:

$$N_{\pm}=\frac{1}{\left[2\left(1\pm S_{AB}\right)\right]^{1/2}} \qquad (8.16)$$

At the equilibrium internuclear separation of 106 pm, the overlap integral is equal to 0.59. Hence, $N_+ = 0.56$ and $N_- = 1.10$. A larger normalization constant is needed for the antibonding orbital because there is destructive interference of the two atomic orbitals between the two nuclei.

At all points the magnitude of $(\psi_A + \psi_B)^2$ must be greater than that of $(\psi_A - \psi_B)^2$, and therefore $\int(\psi_A + \psi_B)^2\,d\tau$ is greater than $\int(\psi_A - \psi_B)^2\,d\tau$. Since $N_+^2\int(\psi_A + \psi_B)^2\,d\tau = N_-^2\int(\psi_A - \psi_B)^2\,d\tau =1$, it follows that N_+ must be less than N_-.

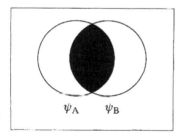

Figure 8.7 The overlap of two 1s atomic wavefunctions. Significant contributions to the overlap integral will occur only in the shaded area

8.7   The Hydrogen Molecule

In this section the molecular orbital and valence bond approaches to bonding in the hydrogen molecule will be compared. In their simplest forms we shall find that valence bond theory is better than MO theory, but as the models become more sophisticated the results obtained by the two methods converge to give the exact experimental result.

8.7.1 The Hamiltonian Operator

The distances between the electrons and nuclei in the hydrogen molecule are defined in Figure 8.8. The electronic Hamiltonian can be written as:

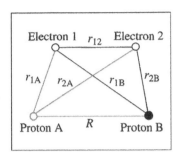

Figure 8.8 Schematic diagram of the hydrogen molecule. The distances of electrons 1 and 2 from the two nuclei are represented by r_{1A}, r_{1B}, and r_{2A}, r_{2B}, respectively. The nuclei are separated by a fixed distance R, and the inter-electron separation is r_{12}

$$\hat{H}_e = -\frac{\hbar^2}{2m_e}\left(\nabla_1^{\,2} + \nabla_2^{\,2}\right)\nabla_e^{\,2} - \frac{e^2}{4\pi\varepsilon_0}\left(\frac{1}{r_{1A}} + \frac{1}{r_{1B}} + \frac{1}{r_{2A}} + \frac{1}{r_{2B}} - \frac{1}{r_{12}}\right) + \frac{e^2}{4\pi\varepsilon_0 R} \qquad (8.17)$$

$\nabla_1^{\,2}$ and $\nabla_2^{\,2}$ are the Laplacian operators for the two electrons. The first term represents the kinetic energy of the two electrons, and the other terms the various electrostatic attractions and repulsions.

8.7.2 The Molecular Orbital (MO) Method

In this approach the extra electron needed to form a hydrogen molecule is added to the $1s\sigma_g$ orbital of H_2^+ in much the same way that a second electron was added to the 1s orbital of atomic hydrogen to form helium. The spatial part of the wavefunction is then $1s\sigma_g(1)1s\sigma_g(2)$, where the number in parentheses indicates the particular electron that is occupying the orbital. According to the Pauli principle, the total wavefunction has to be antisymmetric with respect to interchange of electron co-ordinates. Because the spatial wavefunction is symmetric, the spin component is required to be antisymmetric. The only antisymmetric spin function is $\alpha(1)\beta(2) - \alpha(2)\beta(1)$, in which the electrons have paired spins. The full form of the H_2 ground state wavefunction can therefore be written as:

$$\Phi_{MO}(1,2) = N\left[\psi_A(1) + \psi_B(1)\right]\left[\psi_A(2) + \psi_B(2)\right]\left[\alpha(1)\beta(2) - \alpha(2)\beta(1)\right] \qquad (8.18)$$

where N is the overall normalization constant.

As before, the expectation value for the energy of this trial wavefunction can be calculated by inserting the wavefunction into equation (8.3). This gives:

$$\langle E \rangle = \int \Phi_{MO}^*(1,2)\hat{H}_e\Phi_{MO}(1,2)\,d\tau \qquad (8.19)$$

where $\hat{H}_e$ is now the Hamiltonian operator given in equation (8.17). Integration leads to a rather complicated equation for $\langle E \rangle$ which can be written as:

$$\langle E \rangle = 2E_{1s} + \frac{e^2}{4\pi\varepsilon_0 R} - CI \qquad (8.20)$$

The first term in this equation represents the electronic energy of two hydrogen atoms, and the second term is the electrostatic repulsion between the two nuclei. The term labelled "CI" represents the coulombic interactions of various charge distributions with one another. The integrals can all be evaluated analytically and the potential energy curve obtained. This has a minimum value at an internuclear separation of 84 pm, which corresponds to a dissociation energy of 255 kJ mol^{-1}. These results can be compared with experimental values of 458 kJ mol^{-1} for the dissociation energy and 74.1 pm for the bond length. Clearly, there is considerable room for improvement, and ways in which this can be done will be discussed in a later section.

8.7.3 The Valence Bond (VB) Method

When hydrogen atoms A and B are an infinite distance apart, the electronic wavefunction is given accurately by the equation:

$$\psi_{VB} = \psi_A(1)\psi_B(2) \tag{8.21}$$

Again, the number in parentheses identifies the particular electron occupying the orbital. The two atomic orbitals are multiplied together because the wavefunctions behave rather like probabilities, and the probability of two independent events occurring at the same time is the product of the separate probabilities.

As the atoms are brought closer together they begin to interact with one another, and it is no longer possible to be sure that electron 1 is on atom A and electron 2 on atom B because the detailed trajectories of the electrons cannot be followed with certainty. Therefore, there is an equal probability of finding electron 1 on atom B and electron 2 on atom A. This requires the trial wavefunction for two interacting atoms to be written so that both combinations have equal weighting:

$$\Phi_{VB}(1,2) = \psi_A(1)\psi_B(2) + \psi_A(2)\psi_B(1) \tag{8.22}$$

In order to make the overall wavefunction antisymmetric, this spatial wavefunction has to be multiplied by the antisymmetric spin function, $\alpha(1)\beta(2) - \alpha(2)\beta(1)$. The wavefunction to be used in calculations is therefore:

$$\Phi_{VB}(1,2) = N\left[\psi_A(1)\psi_B(2) + \psi_A(2)\psi_B(1)\right]\left[\alpha(1)\beta(2) - \alpha(2)\beta(1)\right] \tag{8.23}$$

where N is the overall normalization constant.

The expectation value for the energy is then obtained from the equation:

$$\langle E \rangle = \int \Phi_{VB}^*(1,2)\hat{H}_e\Phi_{VB}(1,2)d\tau \tag{8.24}$$

where $\hat{H}_e$ is the Hamiltonian operator given by equation (8.17).

The potential energy curve is found to have a minimum value at $R =$

87 pm, which corresponds to a bond dissociation energy of 303 kJ mol^{-1}. Although the theoretical dissociation energy is still too low, it is closer to the experimental value than the energy obtained from simple MO theory.

8.7.4 Comparison of Methods

The relative success of the valence bond wavefunction comes about because it keeps the two electrons apart. When electron 1 is on atom A, electron 2 is on atom B, and *vice versa*. This can be compared with the MO wavefunction, which assumes that the electrons move independently of one another so that the probability of finding one electron at a particular point in space is independent of the position of the other electron. In reality, of course, electrons tend to avoid one another because they are negatively charged. Thus, the motion of one electron at a particular instant is dependent upon the position of the other electron, and their motions are correlated.

The difference between the two types of wavefunction can be seen by expanding the spatial part of the MO wavefunction in the following way:

$$\Phi_{MO}\left(1,2\right) = N\left[\psi_A\left(1\right) + \psi_B\left(1\right)\right]\left[\psi_A\left(2\right) + \psi_B\left(2\right)\right]$$
$$= N\left[\psi_A\left(1\right)\psi_B\left(2\right) + \psi_A\left(2\right)\psi_B\left(1\right) + \psi_A\left(1\right)\psi_A\left(2\right) + \psi_B\left(1\right)\psi_B\left(2\right)\right] (8.25)$$

The first two terms in this expansion correspond to the VB wavefunction, whereas the last two terms correspond to ionic structures, $H_A^-H_B^+$ and $H_A^+H_B^-$, in which both electrons are on the same atom. The weakness of the MO description is that it gives equal weighting to covalent and ionic structures, and this is at variance with the general perception that the bonding in the hydrogen molecule is mainly covalent.

The VB wavefunction suffers from the opposite problem because it does not allow for any ionic character to be present in the hydrogen bond. This situation can be improved by adding the ionic terms found in the MO wavefunction, but multiplied by a weighting factor λ. With this modification, the spatial part of the VB wavefunction becomes:

$$\Phi_{VB}\left(1,2\right) = N\left\{\psi_A\left(1\right)\psi_B\left(2\right) + \psi_A\left(2\right)\psi_B\left(1\right) + \lambda\left[\psi_A\left(1\right)\psi_A\left(2\right) + \psi_B\left(1\right)\psi_B\left(2\right)\right]\right\} (8.26)$$

When this wavefunction is substituted into equation (8.24), it is found that the energy has a minimum value when the coefficient λ is equal to 1/6. According to the variation principle, this value of λ gives the closest approximation to the true ground state wavefunction. The contribution of ionic structures to the overall bond is equal to λ^2, which has a value of 1/36, or about 3%. This ionic contribution, although small, causes the bond dissociation energy calculated for the VB wavefunction to increase from 303 to 396 kJ mol $^{-1}$. Although this represents a significant improvement over the simple VB theory, the calculated dissociation energy still falls short of the experimental value by 62 kJ mol^{-1}.

Significant improvements to the MO wavefunction can also be made by adding terms which represent the molecular orbitals of higher energy states of the molecule. There is an excited state of the molecule in which both electrons occupy the $1s\sigma_u^*$ antibonding molecular orbital. This can be written as $1s\sigma_u^*(1)1s\sigma_u^*(2)$. It is possible to mix this wavefunction with the one where both electrons are in the bonding molecular orbital because the two wavefunctions have the same symmetry. This results in a considerable increase in the calculated dissociation energy of the molecule. By using even more complex wavefunctions, containing as many as 100 terms,[2] it is possible to obtain results which are so close to the experimental values as to be within experimental error. The energies and bond lengths calculated for the various models are listed in Table 8.1.

Table 8.1 Comparison of binding energies and bond lengths calculated for the hydrogen molecule

Type of wavefunction	Binding energy/kJ mol^{-1}	Bond length/pm
Simple valence bond	303	87
Simple molecular orbital	255	84
Valence bond with ionic contribution	396	74.9
100 term function (see Kolos and Wolniewicz[2])	458	74.1
Experimental result	458	74.1

8.8   Molecular Orbitals for Other Diatomic Molecules

The MO method used with the hydrogen molecule can be applied to other diatomic molecules after some modifications have been made. Each electron is considered to move in the potential field produced by the nuclei, plus some additional electrical field which represents the average effect of all the other electrons. This gives rise to the following one-electron Hamiltonian operator:

$$\hat{H}_e = -\frac{\hbar^2}{2m_e}\nabla_e^2 - \frac{e^2}{4\pi\varepsilon_0}\left(\frac{Z_A}{r_A} + \frac{Z_B}{r_B}\right) + V\left(r_A, r_B\right) \qquad (8.27)$$

In this equation, r_A and r_B are the distances of the electron from the two nuclei, and Z_A and Z_B are the effective charge numbers of the two nuclei after allowance has been made for the screening effect of the other electrons. The final term represents the supposedly static potential field which arises from the charge distribution of the other electrons.

We start by considering the following trial wavefunction for a heteronuclear diatomic molecule, AB:

$$\phi = c_A \psi_A + c_B \psi_B \tag{8.28}$$

Here, ψ_A and ψ_B are atomic orbitals centred on atoms A and B, respectively; c_A and c_B are weighting factors for the two atomic wavefunctions, the best values of which have to be determined. The method to be discussed is quite general and can easily be extended to larger molecules.

The expectation value for the energy of this trial wavefunction can be obtained from equation (8.3):

$$\langle E \rangle = \frac{\int \phi^* \hat{H} \phi \, d\tau}{\int \phi^* \phi \, d\tau} \tag{8.29}$$

It will be assumed that the atomic wavefunctions do not have complex components, and therefore that $\phi^* = \phi$. It is not strictly necessary to do this, but it simplifies the mathematics. When the expression in equation (8.28) is substituted into this equation, we obtain:

$$\langle E \rangle = \frac{\int \left(c_A \psi_A + c_B \psi_B \right) \hat{H} \left(c_A \psi_A + c_B \psi_B \right) d\tau}{\int \left(c_A \psi_A + c_B \psi_B \right)^2 d\tau} \tag{8.30}$$

This can be expanded to give:

$$\langle E \rangle = \frac{c_A^2 H_{AA} + 2 c_A c_B H_{AB} + c_B^2 H_{BB}}{c_A^2 S_{AA} + 2 c_A c_B S_{AB} + c_B^2 S_{BB}} \tag{8.31}$$

In this equation the following symbols have been used to represent the integrals:

$$S_{AA} = \int \psi_A \psi_A d\tau = 1 \quad S_{BB} = \int \psi_B \psi_B \, d\tau = 1$$
$$S_{AB} = \int \psi_A \psi_B d\tau = \int \psi_B \psi_A d\tau = S_{BA}$$
$$H_{AA} = \int \psi_A \hat{H} \psi_A d\tau \quad H_{BB} = \int \psi_B \hat{H} \psi_B d\tau$$
$$H_{AB} = \int \psi_A \hat{H} \psi_B d\tau = \int \psi_B \hat{H} \psi_A d\tau = H_{BA}$$

S_{AA} and S_{BB} will both be equal to one if the atomic orbitals have been normalized. S_{AB} is the overlap integral, already discussed in Section 8.6.

H_{AA} is known as a Coulomb integral. It represents the energy that an electron would have in the diatomic molecule if it occupied the atomic orbital ψ_A. It is more negative than the ground state energy E_A of an electron on atom A alone because of the coulombic attraction of the electron for the second nucleus B. H_{BB} is also a Coulomb integral, centred on atom B.

H_{AB} is known as the resonance integral for reasons that originate in classical physics. It vanishes when the orbitals ψ_A and ψ_B do not overlap. It is normally negative for equilibrium bond lengths.

The next step is to find the minimum energy that can be obtained from equation (8.31) by setting $\partial\langle E\rangle/\partial c_A$ and $\partial\langle E\rangle/\partial c_B$ equal to zero. We start by writing equation (8.31) in the following form:

$$\langle E\rangle\left(c_A^2 + 2c_A c_B S_{AB} + c_B^2\right) = c_A^2 H_{AA} + 2c_A c_B H_{AB} + c_B^2 H_{BB} \qquad (8.32)$$

where use has been made of the fact that S_{AA} and S_{BB} are both equal to one. Differentiation of this equation with respect to the variable c_A gives:

$$\langle E\rangle\left(2c_A + 2c_B S_{AB}\right) + \frac{\partial\langle E\rangle}{\partial c_A}\left(c_A^2 + 2c_A c_B S_{AB} + c_B^2\right) = 2c_A H_{AA} + 2c_B H_{AB} \qquad (8.33)$$

Similarly, differentiation with respect to the variable c_B gives:

$$\langle E\rangle\left(2c_A S_{AB} + 2c_B\right) + \frac{\partial\langle E\rangle}{\partial c_B}\left(c_A^2 + 2c_A c_B S_{AB} + c_B^2\right) = 2c_A H_{AB} + 2c_B H_{BB} \qquad (8.34)$$

At the minimum value of the energy, $\partial\langle E\rangle/\partial c_A$ and $\partial\langle E\rangle/\partial c_B$ are both equal to zero, and equations (8.33) and (8.34) can then be reorganized to give:

$$c_A\left(H_{AA} - E\right) + c_B\left(H_{AB} - ES_{AB}\right) = 0 \qquad (8.35)$$

and

$$c_A\left(H_{AB} - ES_{AB}\right) + c_B\left(H_{BB} - E\right) = 0 \qquad (8.36)$$

It should be noted that we are now using E for the minimum energy rather than $\langle E\rangle$, because this represents the best possible value of the energy that can be obtained with the LCAO method. Equations (8.35) and (8.36) are known as the secular equations. They have non-trivial solutions only if the determinant of the coefficients is equal to zero:

Differentiation is made with respect to the variable c_A, while the other variable, c_B, is treated as a constant.

The trivial solutions to the equations are just $c_A = 0$ and $c_B = 0$.

Although the energy of the antibonding orbital does not correspond to an absolute minimum in the energy, it can be shown that the variation principle is still a valid method of obtaining the best energy for this state.

$$\begin{vmatrix} H_{AA} - E & H_{AB} - ES_{AB} \\ H_{AB} - ES_{AB} & H_{BB} - E \end{vmatrix} = 0 \qquad (8.37)$$

This is known as the secular determinant, and it can be expanded to give a quadratic equation in E. It has two roots, which give the energies of the bonding and antibonding orbitals.

8.9 Molecular Orbitals for Homonuclear Diatomic Molecules

8.9.1 Solutions of the Secular Equations

If atoms A and B are identical, and the molecular orbital is constructed from atomic orbitals of the same type, then $H_{AA} = H_{BB}$, and the secular determinant can be expanded as:

$$\left(H_{AA} - E\right)^2 - \left(H_{AB} - ES_{AB}\right)^2 = 0 \qquad (8.38)$$

This equation has two solutions:

$$E_{\pm} = \frac{H_{AA} \pm H_{AB}}{1 \pm S_{AB}} \qquad (8.39)$$

The overlap integral, S_{AB}, is of the order of 0.2–0.3 for atomic orbitals with $n = 2$, and it is possible to ignore this term in an approximate calculation. It can also be shown that the coulomb integral, H_{AA}, is roughly equal to E_A, the energy of an electron in the atomic orbital ψ_A. With these approximations, equation (8.39) can be written as:

$$E_{\pm} \approx E_A \pm H_{AB} \qquad (8.40)$$

The resonance integral, H_{AB}, is normally negative so that the bonding orbital is associated with the positive sign. The resulting energy level diagram is shown in Figure 8.9a, where it can be seen that the bond energy for the bonding orbital is approximately equal to H_{AB}. In this

Figure 8.9 Energy levels calculated for the bonding and antibonding orbitals of a homonuclear diatomic molecule: (a) overlap integral ignored; (b) overlap integral taken into account. H_{AB} is the resonance integral

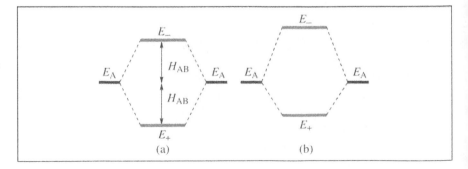

(a)   (b)

diagram the antibonding orbital goes up in energy by the same amount that the bonding orbital goes down in energy, but this is no longer true when the overlap integral is taken into account (see Figure 8.9b).

By symmetry it can be shown that $c_A = \pm c_B$ for a homonuclear diatomic molecule. After normalization, the corresponding molecular orbitals become:

$$\phi_+ = \left\{ \frac{1}{2\left(1 + S_{AB}\right)} \right\}^{1/2} \left\{ \psi_A + \psi_B \right\}$$

(8.41)

and

$$\phi_- = \left\{ \frac{1}{2\left(1 - S_{AB}\right)} \right\}^{1/2} \left\{ \psi_A - \psi_B \right\}$$

(8.42)

8.9.2 Molecular Orbitals for Second-row Homonuclear Diatomics

The atomic orbitals available for bonding in these molecules are 2s and 2p. The 1s orbitals are too compact to give significant overlap, and they are regarded as essentially non-bonding. The overlap of the 2s orbitals results in a bonding $2s\sigma_g$ orbital and an antibonding $2s\sigma_u{}^*$ orbital, which look much like the $1s\sigma_g$ and $1s\sigma_u{}^*$ orbitals shown in Figure 8.5.

The 2p atomic orbitals can interact in two different ways to give the molecular orbitals shown in Figure 8.10. It can be seen that the lobes of the $2p_z$ orbitals are directed along the internuclear axis, and they overlap strongly to give molecular orbitals with the same cylindrical symmetry as the 2s molecular orbitals. Hence a $2p_z\sigma_g$ bonding orbital and a $2p_z\sigma_u{}^*$ antibonding orbital are formed.

The lobes of the $2p_x$ and $2p_y$ orbitals are directed at right angles to the internuclear axis, and they overlap sideways to give either a bonding or an antibonding π orbital. These orbitals have positive and negative lobes running parallel to the internuclear axis, and a node along the axis itself. It can be seen from Figure 8.10b that the bonding orbitals change sign when inverted through the centre of the molecule, which makes them *ungerade*, whereas the antibonding orbitals are *gerade*. We therefore have $2p_x\pi_u$ and $2p_y\pi_u$ bonding orbitals, and $2p_x\pi_g{}^*$ and $2p_y\pi_g{}^*$ antibonding orbitals.

Generally, the $2p_z\sigma_g$ orbital shows stronger overlap than the $2p\pi_u$ orbitals, and it would therefore be expected to have the lower energy. This would result in the MO energy diagram shown in Figure 8.11, which is found to be applicable to oxygen and fluorine. However, spectroscopic evidence shows that the $2p_z\sigma_g$ energy level lies above the $2p\pi_u$ levels for the other second row elements, as shown in Figure 8.12. This change in

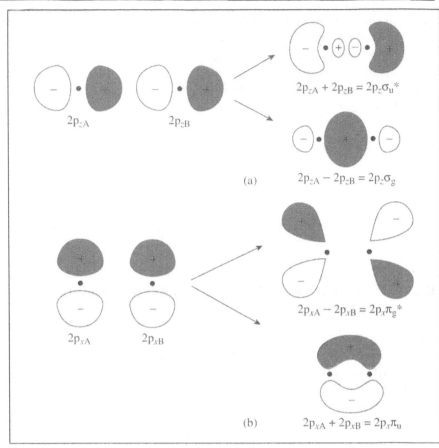

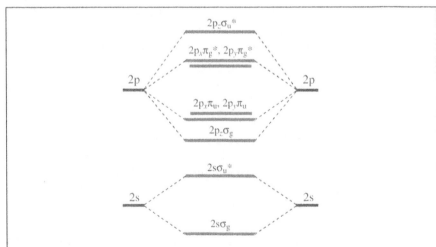

Figure 8.10 The molecular orbitals that can be produced by overlap of 2p atomic orbitals. Part (a) shows the formation of bonding and antibonding σ orbitals from overlap of $2p_z$ atomic orbitals, and part (b) shows the formation of bonding and antibonding π orbitals from overlap of $2p_x$ atomic orbitals

Figure 8.11 Molecular orbital energy diagram in which the $2p_z\sigma_g$ orbital lies below the $2p\pi_u$ orbitals. This arrangement applies to oxygen and fluorine

the ordering of the energy levels occurs because the $2s\sigma_u{}^*$ antibonding orbital and the $2p_z\sigma_g$ bonding orbital have similar energies, and this allows them to mix together to form two new states, one with an energy slightly less than the original $2s\sigma_u^*$ level and the other with an energy slightly higher than the original $2p_z\sigma_g$ level. This mixing is sufficient to push the latter above the $2p\pi_u$ levels. No interaction is possible between the $2s\sigma_u^*$ level and the $2p\pi_u$ levels because they have different symmetries, and therefore zero overlap. The modified molecular orbitals will still be referred to by their original names because the amount of mixing is relatively small.

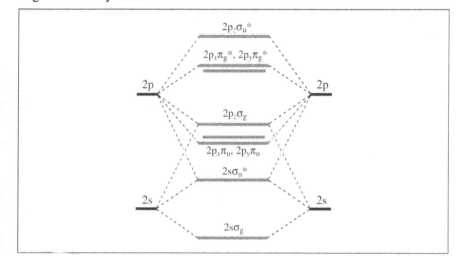

Figure 8.12 Molecular orbital energy diagram in which the $2p_z\sigma_g$ orbital is above the $2p\pi_u$ orbitals. This arrangement applies to nitrogen

The photoelectron spectrum of nitrogen (Figure 1.9) shows three peaks due to emission of electrons from the $2s\sigma_u^*$, $2p\pi_u$ and $2p_z\sigma_g$ orbitals. These have ionization energies of 18.7, 16.7 and 15.6 eV, respectively. Thus, the energy of the $2p_z\sigma_g$ orbital lies above that of the $2p\pi_u$ orbital for nitrogen, as shown in Figure 1.10.

The variation in the energy levels and electron occupancy for homonuclear diatomic molecules of the elements of the second period is shown in Figure 8.13. It can be seen that the energy of the $2p\pi_u$ orbitals remains approximately constant in going from Li_2 to F_2, whereas the energy of the $2p_z\sigma_g$ orbital drops considerably, and finally becomes less than that of the $2p\pi_u$ orbital towards the end of the period.

Figure 8.13 The molecular orbital energy levels of the second-row homonuclear diatomic molecules, and their electron occupancy

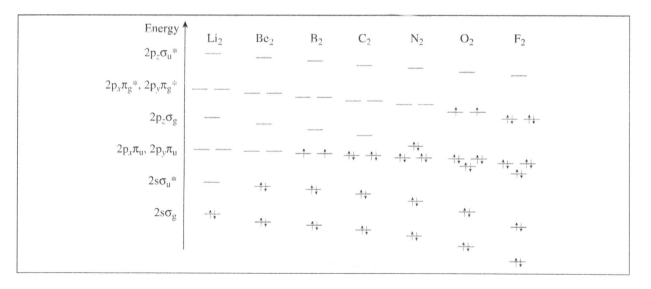

MO theory has been particularly successful in providing an explanation for the paramagnetic nature of the oxygen molecule. It can be seen from Figure 8.13 that oxygen has two unpaired electrons in the degenerate $2p_x\pi_g^*$ and $2p_y\pi_g^*$ orbitals. These electrons occupy separate orbitals because this keeps them far apart, and therefore reduces the coulombic repulsion between them. The electrons have parallel spins in accordance with Hund's rule, and measurements of the paramagnetism of oxygen confirm the presence of two unpaired electrons. Hence, there is complete agreement between theory and experiment.

The strength of the bond formed can be assessed by calculating the bond order according to the following formula:

$$\text{Bond order} = \frac{1}{2}\left[n\big(\text{bonding}\big) - n\big(\text{antibonding}\big)\right] \qquad (8.43)$$

Here, n(bonding) and n(antibonding) are the numbers of electrons in bonding and antibonding orbitals, respectively. The predicted bond orders for the second-row homonuclear diatomics are given in Table 8.2, together with the experimentally measured bond lengths and bond energies. There is seen to be a strong correlation between bond order and bond energy.

Table 8.2 Relationship between bond order, bond length and bond dissociation energy

Molecule	n(bonding)	n(anti-bonding)	Bond order	Bond length/pm	Bond dissociation energy/kJ mol^{-1}
Li_2	2	0	1	267	105
Be_2	2	2	0	245	<10
B_2	4	2	1	159	289
C_2	6	2	2	124	599
N_2	8	2	3	110	942
O_2	8	4	2	121	494
F_2	8	6	1	141	154

Worked Problem 8.2

Q Calculate the bond orders of N_2^+ and N_2^-.

A The electron configuration of N_2^+ will be $(2s\sigma_g)^2(2s\sigma_u^*)^2(2p_x\pi_u)^2$ $(2p_y\pi_u)^2(2p_z\sigma_g)^1$. This has a bond order of $(7-2)/2 = 2.5$. The bond is weaker than that in the neutral molecule because there is one less electron in the $2p_z\sigma_g$ bonding orbital. The ionic species N_2^- has an extra electron in the antibonding $2p\pi_g{}^*$ orbital, and the bond order is therefore $(8 - 3)/2 = 2.5$. Thus, both ionic species are less stable than the neutral molecule.

8.10 Application of MO Theory to Heteronuclear Diatomic Molecules

These are molecules in which different atoms are bonded together, and the atomic orbitals on the two atoms will therefore have different energies. As a general rule, appreciable interaction occurs only when atomic orbitals with broadly similar energies are used to form the molecular orbitals. Two molecules will be used to illustrate the bonding possibilities; the first is nitrogen monoxide (nitric oxide), where the energies of the 2s and 2p orbitals on the carbon and oxygen atoms match up quite well, and the second is hydrogen fluoride, where there is a large disparity between the orbital encrgies on the two atoms.

8.10.1 The NO Molecule

The molecular orbital energy level diagram for NO is shown in Figure 8.14. It is similar to the diagram for molecular oxygen, except that the energy gap between the 2s and 2p orbitals is smaller for nitrogen than it is for oxygen. Molecular orbitals of a given symmetry are numbered in order of increasing energy, and g or u subscripts are not used because the molecule does not have a centre of symmetry. The diagram is slightly oversimplified in that there will be some 2p character in the 1σ and $2\sigma^*$

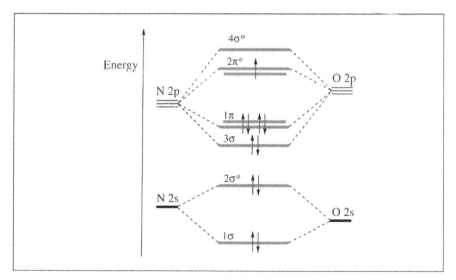

Figure 8.14 The molecular orbital energy level diagram for nitrogen monoxide

orbitals, and some 2s character in the 3σ and $4\sigma^*$ orbitals, but the effect will be relatively small.

The number of valence electrons that need to be accommodated in the molecular orbitals is eleven, five coming from the nitrogen atom and six from the oxygen atom. Pairs of electrons occupy the molecular orbitals in order of increasing energy until the antibonding $2\pi^*$ orbitals are reached, which are occupied by a single unpaired electron. There are eight electrons in bonding orbitals and three in antibonding orbitals, and the bond order is therefore 2.5. The unpaired electron causes the molecule to be paramagnetic and highly reactive.

8.10.2 The HF Molecule

The relative energies of the atomic and molecular orbitals of this polar molecule are shown in Figure 8.15, where it can be seen that the fluorine 1s and 2s atomic orbitals lie far below the hydrogen 1s orbital on the energy scale. As a consequence, they generate molecular orbitals which are essentially unmodified fluorine atomic orbitals, with hardly any contribution from the hydrogen atom. Although the fluorine 2p orbitals have energies that are sufficiently close to that of the hydrogen 1s atomic orbital for interaction to be feasible, only the $2p_z$ orbital has the correct symmetry for bonding to occur. The bonding molecular orbital created by the $F(2p_z)$–$H(1s)$ overlap is shown in Figure 8.16. The wavefunction can be written as:

$$\phi\left(\text{bonding}\right) = 0.33\psi_{1s}\left(H\right) + 0.94\psi_{2p_z}\left(F\right) \tag{8.44}$$

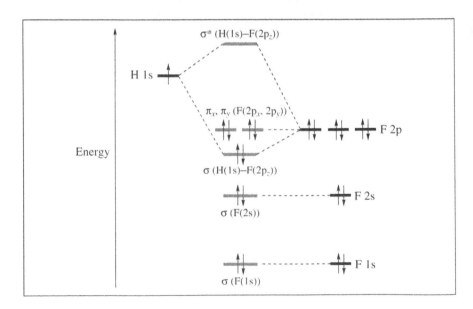

Figure 8.15 Energy level diagram for hydrogen fluoride

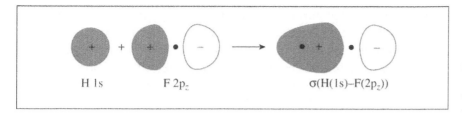

Figure 8.16 The bonding molecular orbital formed from the overlap of the H 1s and the F $2p_z$ atomic orbitals

This shows that the major contribution comes from the fluorine atom, and the electrons in the bonding orbital are therefore most likely to be found close to the fluorine atom, which means that the molecule is highly polarized. There is also an antibonding orbital with the wavefunction:

$$\phi\left(\text{antibonding}\right) = 0.94\psi_{1s}\left(H\right) - 0.33\psi_{2p_z}\left(F\right) \qquad (8.45)$$

Here, the major contribution comes from the hydrogen atom.

It can be seen from Figure 8.17 that the $2p_x$ and $2p_y$ orbitals produce zero net overlap with the hydrogen 1s orbital because the overlap of the positive lobe cancels out the overlap of the negative lobe. Therefore, the hydrogen 1s orbital makes no contribution to the π molecular orbitals.

In this simple treatment it has been assumed that the fluorine atomic orbitals do not mix, which is not strictly true. More detailed calculations show that the F 2s orbital makes a significant contribution to the bonding orbital.

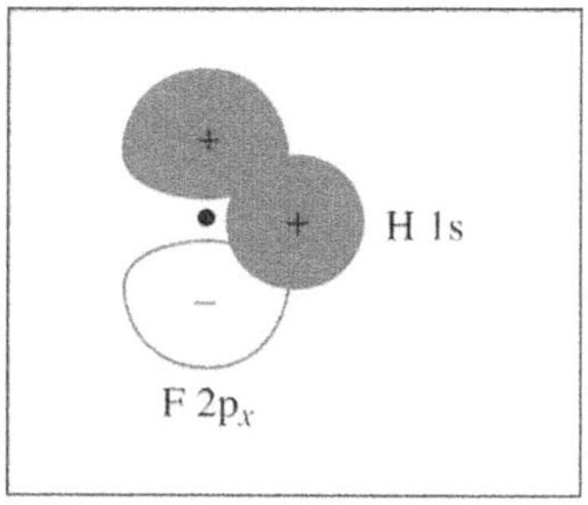

Figure 8.17 Illustration of the zero net overlap produced by combining the H 1s and F $2p_x$ orbitals

8.11 Hybridization in Polyatomic Molecules

We have already seen that more than one atomic orbital on an atom may contribute to a particular molecular orbital. For instance, the ordering of the molecular orbitals for homonuclear diatomic molecules depended upon the mixing of 2s and 2p atomic orbitals. A similar mixing is found in the bonding of HF. One disadvantage of this approach is that we appear to have lost the simple concept of a bond as an overlap of two atomic orbitals, one on each atom. Often it is useful to work with hybrid orbitals, which consist of a linear combination of the atomic orbitals on a single atom. These hybrid orbitals are then thought of as overlapping with orbitals located on other atoms to form chemical bonds. This is essentially the VB approach to bonding, in which bonds are seen as localized between pairs of atoms.

It should be emphasized that hybrid orbitals are not true solutions of the Schrödinger equation for the atom concerned, and they are not generally used in rigorous quantum theory. They are, however, often extremely useful in gaining insight into the shapes of polyatomic molecules. We will examine their use in predicting the shapes of the molecules BeH_2, BH_3, CH_4, NH_3 and H_2O.

The electron configuration of the ground state of the beryllium atom is $1s^2 2s^2$, and it might be thought that bonding in BeH_2 would occur by overlap between the hydrogen 1s orbital and the beryllium 2s orbital. In fact, there is a significant contribution from one of the 2p orbitals. Bond formation can be thought of as occurring in three stages:

(i)   A 2s electron on the beryllium atom is promoted into a 2p orbital, which will be taken to be the $2p_z$ orbital.
(ii) The 2s and $2p_z$ orbitals then combine to form two hybrid sp orbitals, each occupied by an electron, as shown in Figure 8.18. These hybrid orbitals have one lobe greater than the other because there is constructive overlap of the 2s and $2p_z$ orbitals on one side of the nucleus and destructive overlap on the other. The normalized orbitals formed in this way are

$$\phi_{sp+} = 2^{-1/2}\left(\psi_{2s} + \psi_{2p_z}\right) \tag{8.46}$$

$$\phi_{sp-} = 2^{-1/2}\left(\psi_{2s} - \psi_{2p_z}\right) \tag{8.47}$$

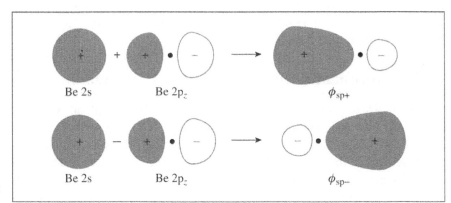

Figure 8.18 The formation of two sp hybrid orbitals by combining the 2s and $2p_z$ atomic orbitals of a beryllium atom

(iii) Finally, the major lobes of the two hybrid sp orbitals overlap with the 1s atomic orbitals of the hydrogen atoms to form the following bonding orbitals:

$$\phi_A = c_1 \psi_{1s}\left(H_A\right) + c_2 \psi_{sp+}\left(Be\right) \tag{8.48}$$

$$\phi_B = c_1 \psi_{1s}\left(H_B\right) + c_2 \psi_{sp-}\left(Be\right) \tag{8.49}$$

where c_1 and c_2 are coefficients, the values of which have to be determined.

Because the major lobes of the hybrid orbitals are much more directed than the 2s orbital, a much stronger bond can be formed, and the extra energy released is more than sufficient to compensate for the energy required to promote an electron from the 2s to the 2p atomic orbital of beryllium. From the orientation of the two hybrid orbitals, the molecule is expected to be linear with a H–Be–H bond angle of 180°.

Next we turn to the BH_3 molecule, in which the three B–H bonds are coplanar, and the H–B–H angle is 120°. The electronic configuration of the boron atom in the ground state is $1s^2 2s^2 2p$, and hybridization can occur only after promotion of an electron from the 2s orbital to a 2p orbital to give a $1s^2 2s^1 2p^2$ configuration. Three equivalent hybrid sp^2 orbitals can then be constructed as follows:

$$\phi_1 = \frac{1}{\sqrt{3}}\psi_{2s} + \sqrt{\frac{2}{3}}\psi_{2p_x} \qquad (8.50)$$

$$\phi_2 = \frac{1}{\sqrt{3}}\psi_{2s} - \frac{1}{\sqrt{6}}\psi_{2p_x} + \frac{1}{\sqrt{2}}\psi_{2p_y} \qquad (8.51)$$

$$\phi_3 = \frac{1}{\sqrt{3}}\psi_{2s} - \frac{1}{\sqrt{6}}\psi_{2p_x} - \frac{1}{\sqrt{2}}\psi_{2p_y} \qquad (8.52)$$

These orbitals lie in a plane, as shown in Figure 8.19, and they overlap with the hydrogen 1s orbitals to give three equivalent B–H bonds. The $2p_z$ orbital lies at right angles to this plane and takes no part in bond formation.

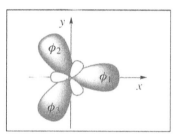

Figure 8.19 The three sp^2 hybrid orbitals which can be formed on a boron atom from a combination of the 2s, $2p_x$ and $2p_y$ atomic orbitals

Worked Problem 8.3

Q Show that the hybrid orbitals given in equations (8.50)–(8.52) are normalized.

A Consider the hybrid orbital:

$$\phi = c_1\psi_{2s} + c_2\psi_{2p_x} + c_3\psi_{2p_y} \qquad (8.53)$$

where the atomic wavefunctions ψ_{2s}, ψ_{2p_x} and ψ_{2p_y} are all real and normalized. This hybrid orbital will itself be normalized if:

$$\int \phi^2 d\tau = \int \left(c_1\psi_{2s} + c_2\psi_{2p_x} + c_3\psi_{2p_y}\right)^2 d\tau = 1 \qquad (8.54)$$

A set of atomic wavefunctions, such as ψ_{2s}, ψ_{2p_x} and ψ_{2p_y}, have the important property that their overlap integrals are all equal to zero; that is:

$$\int \psi_{2s} \psi_{2p_x} d\tau = \int \psi_{2s} \psi_{2p_y} d\tau = \int \psi_{2p_x} \psi_{2p_y} d\tau = 0 \qquad (8.55)$$

When this condition applies, the wavefunctions are said to be orthogonal to one another. Equation (8.54) then reduces to:

$$\int \phi^2 d\tau = c_1^2 \int \psi_{2s}^2 d\tau + c_2^2 \int \psi_{2p_x}^2 d\tau + c_3^2 \int \psi_{2p_y}^2 d\tau = c_1^2 + c_2^2 + c_3^2 = 1 \qquad (8.56)$$

Examination of equations (8.50)–(8.52) shows that the sum of the squares of the coefficients is in each case equal to one, so that the hybrid orbitals are all normalized.

It is generally true that there is zero net overlap between any two eigenfunctions, ψ_n and ψ_m, of a particular Hamiltonian operator, ψ_n and ψ_m. Thus, $\int \psi_n \psi_m d\tau = 0$.

The best known example of hybridization occurs with the carbon atom, where the 2s orbital and the three 2p orbitals can be combined to give four sp^3 hybrid orbitals. The ground state of the carbon atom has the electronic configuration $1s^2 2s^2 2p^2$, and this needs to be raised to $1s^2 2s^1 2p^3$ before hybridization can occur. The sp^3 hybrid orbitals are:

$$\phi_1 = \frac{1}{2} \left(\psi_{2s} + \psi_{2p_x} + \psi_{2p_y} + \psi_{2p_z} \right) \qquad (8.57)$$

$$\phi_2 = \frac{1}{2} \left(\psi_{2s} - \psi_{2p_x} - \psi_{2p_y} + \psi_{2p_z} \right) \qquad (8.58)$$

$$\phi_3 = \frac{1}{2} \left(\psi_{2s} + \psi_{2p_x} - \psi_{2p_y} - \psi_{2p_z} \right) \qquad (8.59)$$

$$\phi_4 = \frac{1}{2} \left(\psi_{2s} - \psi_{2p_x} + \psi_{2p_y} - \psi_{2p_z} \right) \qquad (8.60)$$

These orbitals form the tetrahedral structure seen in Figure 8.20. There is a single electron in each hybrid orbital, and these are available for bonding to four hydrogen atoms to make CH_4.

We find a similar form of sp^3 hybridization in ammonia. The nitrogen atom has the electronic configuration $1s^2 2s^2 2p_x^1 2p_y^1 2p_z^1$, and there are now five electrons to be accommodated in the four hybrid orbitals. Thus, one hybrid orbital must contain two electrons from the nitrogen atom with their spins paired. These are known as a non-bonding pair or

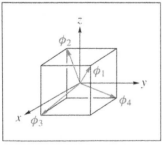

Figure 8.20 The four tetrahedral sp^3 hybrid orbitals which can be formed on a carbon atom by combination of the 2s orbital with the three 2p orbitals

lone pair, and they will be relatively unreactive. The other three electrons are available for bonding to three hydrogen atoms to form NH_3. On this basis, ammonia would be expected to have the structure shown in Figure 8.21, in which the H atoms are positioned at three corners of a tetrahedron and there is a non-bonding pair of electrons located at the fourth corner. According to this model, the H–N–H bond angle should be 109°, which is in reasonably close proximity to the experimentally determined value of 107°. The reason for the slight discrepancy is that the four hybrid orbitals are not equivalent: three of them are involved in bonding to hydrogen atoms, and the fourth contains the non-bonding pair of electrons. This results in the 2p contribution to the bonding hybrid orbitals being slightly greater than that for the non-bonding orbital, with a concomitant small decrease in the bond angle.

8.12 Bonding in the Water Molecule

8.12.1 Valence Bond View

The shape of the water molecule can be explained by hybridization of the 2s, $2p_y$ and $2p_z$ atomic orbitals on the oxygen atom, as shown in Figure 8.22. The $2p_x$ atomic orbital points in a direction at right angles to the plane of the molecule, and is not involved in bonding. The three hybrid orbitals, consistent with a bond angle of 104.5°, are:

$$\phi_1 = 0.45\psi_{2s} + 0.71\psi_{2p_y} + 0.55\psi_{2p_z} \tag{8.61}$$

$$\phi_2 = 0.45\psi_{2s} - 0.71\psi_{2p_y} + 0.55\psi_{2p_z} \tag{8.62}$$

$$\phi_3 = 0.77\psi_{2s} - 0.63\psi_{2p_z} \tag{8.63}$$

The first two hybrid orbitals point towards the hydrogen atoms, and they differ only in the sign of the contribution from the $2p_y$ orbital. The third hybrid orbital does not include a contribution from the $2p_y$ orbital, and is therefore aligned along the z axis. It points away from the two hydrogen atoms, which makes it non-bonding. There are four electrons on the oxygen atom which have to be accommodated in these orbitals; one goes into each bonding orbital and the other two go into the non-bonding orbital.

The H–O–H bond angle is determined by the proportions of the $2p_y$ and $2p_z$ orbitals that contribute to the hybrid orbitals. These may be represented as vectors with lengths proportional to the coefficients 0.71

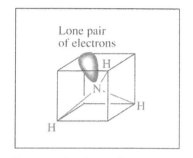

Figure 8.21 The sp³ hybrid orbitals of ammonia. Three of the electrons from the nitrogen atom form bonds with hydrogen atoms, but there are two electrons left over, and these form a non-bonding pair

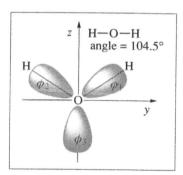

Figure 8.22 Orientation of the water molecule with respect to the coordinate axes

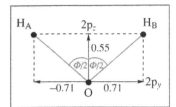

Figure 8.23 Vector diagram relating the H–O–H bond angle to the relative contributions made by the $2p_y$ and $2p_z$ orbitals to the bonding

and 0.55. It can be seen from Figure 8.23 that the bond angle, Φ, is obtained from the formula: $\tan(\Phi/2) = 0.71/0.55 = 1.29$. This gives $\Phi/2 = 52.24°$, which corresponds to a bond angle of 104.5°, as required.

Worked Problem 8.4

Q What form would the hybrid orbitals take if the H–O–H bond angle, Φ, were (a) 90° and (b) 180°?

A The general forms taken by the three hybrid orbitals are:

$$\phi_1 = c_1\psi_{2s} + c_2\psi_{2p_y} + c_3\psi_{2p_z}$$

$$\phi_2 = c_1\psi_{2s} - c_2\psi_{2p_y} + c_3\psi_{2p_z}$$

$$\phi_3 = c_4\psi_{2s} - c_5\psi_{2p_z}$$

Normalization of the hybrid orbitals requires that the sum of the squares of the coefficients must be equal to one: $c_1^2 + c_2^2 + c_3^2 = c_4^2 + c_5^2 = 1$. Similarly, the fractional contributions made by a particular atomic orbital to the hybrid orbitals must add up to one. Thus, $2c_1^2 + c_4^2 = 2c_2^2 = 2c_3^2 + c_5^2 = 1$. It follows that $c_2 = 1/\sqrt{2} = 0.707$, irrespective of the value of the angle Φ.

(a) With $\Phi/2 = 45°$, the vector diagram gives $c_3 = c_2 = 0.707$. Normalization of ϕ_1 then requires that c_1 be zero, showing that the 2s orbital does not contribute to the bonding hybrid orbitals. At this angle the ϕ_3 orbital is a pure 2s orbital.

(b) With $\Phi/2 = 90°$, the vector diagram gives $c_3 = 0$. Normalization of ϕ_1 then requires that $c_1 = 0.707 = c_2$, showing that the hybrid orbitals contain equal contributions from the 2s and $2p_y$ orbitals. At this angle the non-bonding hybrid orbital is a pure $2p_z$ orbital.

It is clear from these calculations that the fractional 2s character of the bonding orbitals increases with the H–O–H bond angle, being zero at 90° and reaching a maximum value of 0.50 at 180°.

8.12.2 Molecular Orbital Picture

The 2s, $2p_x$, $2p_y$ and $2p_z$ atomic orbitals on the oxygen atom combine with the two hydrogen 1s atomic orbitals to give the six molecular orbitals shown in Figure 8.24. The names given to the orbitals are obtained from group theory, but they will be used here merely as labels. Where two or more molecular orbitals have the same symmetry they are numbered consecutively, starting with the one with the lowest energy.

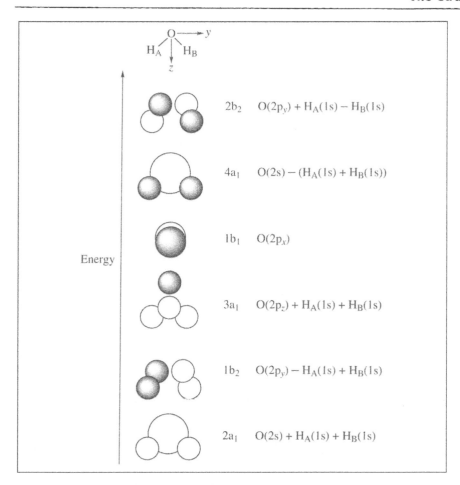

Figure 8.24 Molecular orbital description of the water molecule. The diagram shows the six molecular orbitals that can be formed from the hydrogen 1s and the oxygen 2s and 2p orbitals

The positions of the energy levels with respect to the atomic orbitals are shown in Figure 8.25. The oxygen atom contributes six electrons, and the hydrogen atoms one each, so that there are eight electrons to occupy the molecular orbitals. These fill the four lowest levels, as shown.

There are two quite strongly bonding orbitals, $2a_1$ and $1b_2$, which come about by overlap of the O 2s and O $2p_y$ orbitals with the hydrogen 1s orbitals. However, the $2a_1$ orbital has some $2p_z$ character because it is allowed to mix with the $3a_1$ orbital, which has the same symmetry. By the same reasoning, the $3a_1$ orbital has some 2s character.

The $3a_1$ orbital arises because of the overlap of the O $2p_z$ orbital with the hydrogen 1s orbitals. This is relatively weak, and this orbital is therefore close to being non-bonding. The $1b_1$ orbital consists of an O $2p_x$ orbital, which is directed at right angles to the plane of the molecule, and is therefore non-bonding.

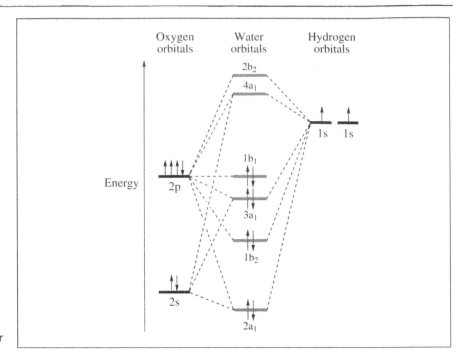

Figure 8.25 The relative energies of the atomic and molecular orbitals of the water molecule

8.13 Hückel Molecular Orbital Theory

The bonding in conjugated organic molecules, such as butadiene, was described in Chapter 2 as a mixture of localized σ bonding, and delocalized π bonding. The σ electrons were considered to form the rigid bonds that determined the shape of the molecule, whilst the π electrons were free to move around the whole molecule in a constant potential field. Although this free π electron model had some success in explaining certain features of the absorption spectra of conjugated molecules, the assumption that the π electrons moved in a constant potential field was obviously a poor approximation. In this section we shall take this simple model a little further by constructing delocalized π molecular orbitals that take account of the undulating nature of the potential energy. The treatment that we shall follow was first introduced by Erich Hückel in 1930.

8.13.1 Application to Butadiene

The two types of bonding found in conjugated molecules are illustrated for butadiene in Figures 2.6 and 2.7. Carbon–carbon σ bonds are formed by overlap of sp^2 hybrid orbitals on the four carbon atoms, and further σ bonds are formed between the carbon hybrid sp^2 orbitals and the hydrogen 1s orbitals. These orbitals make up the σ-bonded framework of the molecule and they all lie in the same plane (see Figure 2.6). The $2p_z$ atomic orbitals on the carbon atoms are oriented at right angles to this plane, and they overlap to form a π molecular orbital, as shown in Figure 2.7. Because of their different symmetry, the $2p_z$ orbitals do not interact with the σ-bonded framework, and they are treated quite separately in Hückel theory.

The molecular orbital for the π electrons can be written as a linear combination of the $2p_z$ atomic orbitals on the four carbon atoms, which we shall denote as ψ_1, ψ_2, ψ_3 and ψ_4. The trial wavefunction therefore becomes:

$$\phi = c_1\psi_1 + c_2\psi_2 + c_3\psi_3 + c_4\psi_4 \tag{8.64}$$

where c_1–c_4 are the weighting coefficients. The variation principle is then used to find the optimum values for these coefficients by minimizing the energies, as discussed in Section 8.8. This leads to the following set of secular equations:

$$c_1\left(H_{11} - ES_{11}\right) + c_2\left(H_{12} - ES_{12}\right) + c_3\left(H_{13} - ES_{13}\right) + c_4\left(H_{14} - ES_{14}\right) = 0 \tag{8.65}$$

$$c_1\left(H_{21} - ES_{21}\right) + c_2\left(H_{22} - ES_{22}\right) + c_3\left(H_{23} - ES_{23}\right) + c_4\left(H_{24} - ES_{24}\right) = 0 \tag{8.66}$$

$$c_1\left(H_{31} - ES_{31}\right) + c_2\left(H_{32} - ES_{32}\right) + c_3\left(H_{33} - ES_{33}\right) + c_4\left(H_{34} - ES_{34}\right) = 0 \tag{8.67}$$

$$c_1\left(H_{41} - ES_{41}\right) + c_2\left(H_{42} - ES_{42}\right) + c_3\left(H_{43} - ES_{43}\right) + c_4\left(H_{44} - ES_{44}\right) = 0 \tag{8.68}$$

These have non-trivial solutions only when the secular determinant is equal to zero:

$$\begin{vmatrix} H_{11} - ES_{11} & H_{12} - ES_{12} & H_{13} - ES_{13} & H_{14} - ES_{14} \\ H_{21} - ES_{21} & H_{22} - ES_{22} & H_{23} - ES_{23} & H_{24} - ES_{24} \\ H_{31} - ES_{31} & H_{32} - ES_{32} & H_{33} - ES_{33} & H_{34} - ES_{34} \\ H_{41} - ES_{41} & H_{42} - ES_{42} & H_{43} - ES_{43} & H_{44} - ES_{44} \end{vmatrix} = 0 \tag{8.69}$$

This results in a quartic equation in the energy E, which has four roots, and is not easy to solve. However, the equation can be greatly simplified by making the following Hückel approximations:

(i) All overlap integrals, S_{12}, S_{13}, etc., are set equal to zero.
(ii) All Coulomb integrals, H_{11}, H_{22}, etc., are given the value α.
(iii) All resonance integrals between neighbouring carbon atoms, H_{12},

H_{23}, etc., are given the value β.

(iv) All resonance integrals between non-neighbouring carbon atoms, H_{13}, H_{14}, etc., are set equal to zero.

Although (iv) is a reasonable approximation because atomic orbitals on non-neighbouring carbon atoms do not overlap significantly, the other approximations all introduce significant error into the calculation. Nevertheless, the method is widely used to give a qualitative description of the bonding in conjugated hydrocarbon molecules.

With these approximations, and assuming that the atomic $2p_z$ orbitals are normalized so that S_{11}, S_{22}, etc. are equal to one, the determinant becomes:

$$\begin{vmatrix} \alpha - E & \beta & 0 & 0 \\ \beta & \alpha - E & \beta & 0 \\ 0 & \beta & \alpha - E & \beta \\ 0 & 0 & \beta & \alpha - E \end{vmatrix} = 0 \qquad (8.70)$$

This can be further simplified by dividing each row by β, and writing $x = (\alpha - E)/\beta$.

The determinant then becomes:

$$\begin{vmatrix} x & 1 & 0 & 0 \\ 1 & x & 1 & 0 \\ 0 & 1 & x & 1 \\ 0 & 0 & 1 & x \end{vmatrix} = 0 \qquad (8.71)$$

This form of the determinant can then be expanded as follows:

$$\text{Determinant} = x \begin{vmatrix} x & 1 & 0 \\ 1 & x & 1 \\ 0 & 1 & x \end{vmatrix} - \begin{vmatrix} 1 & 1 & 0 \\ 0 & x & 1 \\ 0 & 1 & x \end{vmatrix}$$

$$= x^2 \begin{vmatrix} x & 1 \\ 1 & x \end{vmatrix} - x \begin{vmatrix} 1 & 1 \\ 0 & x \end{vmatrix} - \begin{vmatrix} x & 1 \\ 1 & x \end{vmatrix} + \begin{vmatrix} 0 & 1 \\ 0 & x \end{vmatrix}$$

$$= x^2 \left(x^2 - 1 \right) - x^2 - \left(x^2 - 1 \right) + 0$$

$$= x^4 - 3x^2 + 1 = 0 \qquad (8.72)$$

This is a quadratic equation in x^2 with the following roots: $x^2 = 2.62$ and 0.382. The allowed values of x are ± 1.62 and ± 0.62, which lead to the following energies for the four π molecular orbitals:

$$E_1 = \alpha + 1.62\beta, \quad E_2 = \alpha + 0.62\beta$$
$$E_3 = \alpha - 0.62\beta, \quad E_4 = \alpha - 1.62\beta$$

Since β is negative, the lowest energy is $\alpha + 1.62\beta$. The four π electrons in butadiene occupy the two lowest lying states with their spins paired, as shown in Figure 8.26.

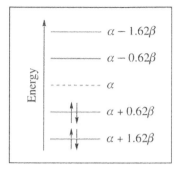

Figure 8.26 The energy levels occupied by the π electrons in the ground state of butadiene

Worked Problem 8.5

Q Calculate values for the coefficients c_1, c_2, c_3 and c_4 for the Hückel molecular orbital with the lowest energy.

A When the Hückel approximations are introduced, the secular equations in (8.65)–(8.68) become:

$$c_1\left(\alpha - E\right) + c_2\beta \qquad + 0 \qquad + 0 \qquad = 0$$
$$c_1\beta \qquad + c_2\left(\alpha - E\right) + c_3\beta \qquad + 0 \qquad = 0$$
$$0 \qquad + c_2\beta \qquad + c_3\left(\alpha - E\right) + c_4\beta \qquad = 0$$
$$0 \qquad + 0 \qquad + c_3\beta \qquad + c_4\left(\alpha - E\right) = 0$$

After dividing each equation by β, and putting $x = (\alpha - E)/\beta$, these equations simplify to:

$$c_1 x + c_2 = 0 \qquad \text{(i)}$$
$$c_1 + c_2 x + c_3 = 0 \qquad \text{(ii)}$$
$$c_2 + c_3 x + c_4 = 0 \qquad \text{(iii)}$$
$$c_3 + c_4 x = 0 \qquad \text{(iv)}$$

The constant, c_1, can be eliminated from equations (i) and (ii) to give:

$$c_2\left(x - \frac{1}{x}\right) + c_3 = 0 \qquad \text{(v)}$$

Putting $x = -1.62$ for the wavefunction with the lowest energy, we find that the term in parentheses is equal to -1. Thus, c_2 must equal c_3. This result can then be combined with equations (i) and (iv) to show that $c_1 = c_4$. From (i) we also have $c_2 = 1.62c_1$. It is therefore possible to express all the constants in terms of c_1. To obtain the numerical coefficients we have to normalize the wavefunction. This requires that:

$$c_1^2 + c_2^2 + c_3^2 + c_4^2 = c_1^2 + \left(1.62c_1\right)^2 + \left(1.62c_1\right)^2 + c_1^2 = 1$$

This gives $c_1 = 0.37$, and the wavefunction becomes:

$$\phi_1 = 0.37\psi_1 + 0.60\psi_2 + 0.60\psi_3 + 0.37\psi_4 \qquad (8.73)$$

This wavefunction is illustrated in Figure 8.27.

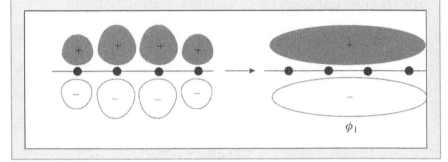

Figure 8.27 Formation of the ϕ_1 Hückel molecular orbital of butadiene from the $2p_z$ atomic orbitals. The black dots represent the carbon nuclei

The other wavefunctions can be obtained in the same way. They are:

$$\phi_2 = 0.60\psi_1 + 0.37\psi_2 - 0.37\psi_3 - 0.60\psi_4 \qquad (8.74)$$

$$\phi_3 = 0.60\psi_1 - 0.37\psi_2 - 0.37\psi_3 + 0.60\psi_4 \qquad (8.75)$$

$$\phi_4 = 0.37\psi_1 - 0.60\psi_2 + 0.60\psi_3 - 0.37\psi_4 \qquad (8.76)$$

These are illustrated in Figure 8.28. The dotted lines follow the general contours of the wavefunctions, and it can be seen that these show some similarity with the one-dimensional particle-in-a-box wavefunctions, discussed in Chapter 2.

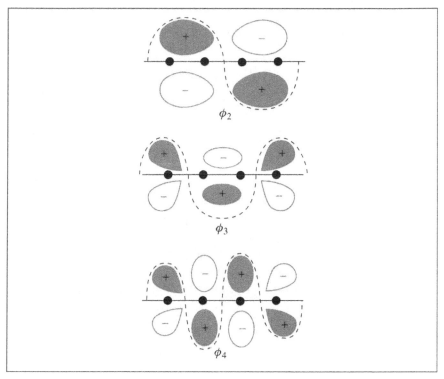

Figure 8.28 The ϕ_2, ϕ_3 and ϕ_4 Hückel molecular orbitals of butadiene

Worked Problem 8.6

Q Derive expressions for the wavefunctions and energies of the Hückel molecular orbitals of ethene.

A The LCAO wavefunctions for ethene are given by the equation:

$$\phi = c_1\psi_1 + c_2\psi_2 \qquad (8.77)$$

The expectation values for the energy are then obtained from the secular determinant:

$$\begin{vmatrix} H_{11} - ES_{11} & H_{12} - ES_{12} \\ H_{21} - ES_{21} & H_{22} - ES_{22} \end{vmatrix} = 0$$

After application of the Hückel approximations, the determinant becomes:

$$\begin{vmatrix} \alpha - E & \beta \\ \beta & \alpha - E \end{vmatrix} = 0$$

This yields the quadratic equation:

$$\left(\alpha - E\right)^2 = \beta^2$$

and the energies are

$$E_{\pm} = \alpha \pm \beta \qquad (8.78)$$

The coefficients c_1 and c_2 in equation (8.77) are obtained from the secular equations:

$$c_1\left(\alpha - E\right) + c_2\beta = 0$$
$$c_1\beta + c_2\left(\alpha - E\right) = 0$$

These give $c_1 = c_2$ for the ground state, with $E = \alpha + \beta$, and $c_1 = -c_2$ for the antibonding state, with $E = \alpha - \beta$. The shape of the corresponding orbitals are shown in Figure 8.29.

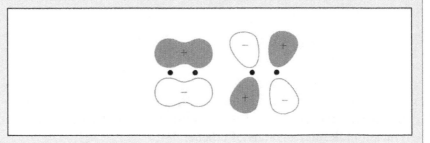

Figure 8.29 The bonding and antibonding Hückel molecular orbitals of ethene

8.13.2 Delocalization Energy

We can calculate the total π electron binding energy in butadiene by summing the energies of each π electron. There are two electrons in the ground state with energy $(\alpha + 1.62\beta)$, and two electrons in the next lowest state with energy $(\alpha + 0.62\beta)$. Thus, the total π electron energy in butadiene is given by the equation:

$$E = 2\left(\alpha + 1.62\beta\right) + 2\left(\alpha + 0.62\beta\right) = 4\alpha + 4.48\beta \qquad (8.79)$$

This can be compared with the π electron energy of two ethene molecules, which is $2 \times 2(\alpha + \beta) = 4\alpha + 4\beta$. We can see that the π electron energy of butadiene is lower than that of two localized π bonds by an amount equal to -0.48β (remember that β is a negative quantity). This is equivalent to about 36 kJ mol⁻¹. The lowering of the energy is due to the ability of π electrons in conjugated systems to move along the full length of the molecule, and it is known as the delocalization energy

8.13.3 Application to Aromatic Systems

The stability of aromatic molecules can be related to the delocalization of the π electrons in the carbon rings. Benzene, which provides the best example, will be discussed here. Hybridization of the atomic orbitals on the carbon atoms produces sp² hybrid orbitals which overlap to give σ bonding within the plane of the hexagonal molecule, as shown in Figure 8.30. The carbon 2p$_z$ orbitals have their lobes at right angles to this plane, and there is therefore no net overlap between these orbitals and the σ bonded framework of the molecule. They form a separate molecular orbital, which is illustrated in Figure 8.31.

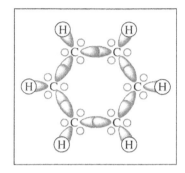

Figure 8.30 The overlap of the hybrid sp² orbitals in benzene

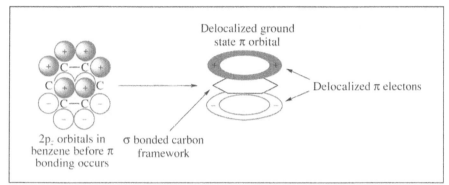

Figure 8.31 The overlap and delocalization of the 2p$_z$ orbitals in benzene

Numbering the carbon atoms from 1 to 6, we can write the Hückel molecular orbitals as:

$$\phi = c_1\psi_1 + c_2\psi_2 + c_3\psi_3 + c_4\psi_4 + c_5\psi_5 + c_6\psi_6 \qquad (8.80)$$

where ψ_1, ψ_2, etc., are 2p$_z$ atomic orbitals on the carbon atoms.

The energies of the Hückel molecular orbitals are obtained by solving the secular determinant:

$$\begin{vmatrix} \alpha - E & \beta & 0 & 0 & 0 & \beta \\ \beta & \alpha - E & \beta & 0 & 0 & 0 \\ 0 & \beta & \alpha - E & \beta & 0 & 0 \\ 0 & 0 & \beta & \alpha - E & \beta & 0 \\ 0 & 0 & 0 & \beta & \alpha - E & \beta \\ \beta & 0 & 0 & 0 & \beta & \alpha - E \end{vmatrix} = 0 \qquad (8.81)$$

This is a simple extension of the secular determinant for butadiene, except that an extra β now appears at the end of the first row, and another one at the beginning of the last row. These represent the terms H_{16} and H_{61}, respectively. For a linear molecule these terms would be put equal to zero because carbon atoms 1 and 6 would be a long way from one another, but in the benzene ring these carbon atoms are nearest neighbours.

Expansion of the secular determinant results in an algebraic equation with six roots. These are:

$$E = \alpha \pm 2\beta, \quad \alpha \pm \beta, \quad \alpha \pm \beta \quad\quad (8.82)$$

The lowest energy is $\alpha + 2\beta$ and the highest energy is $\alpha - 2\beta$. In between, there are two doubly degenerate states with energies of $\alpha + \beta$ and $\alpha - \beta$. The energy levels, and their associated molecular orbitals, are shown in Figure 8.32.

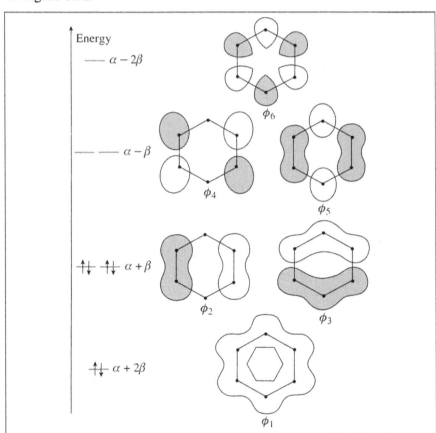

Figure 8.32 The Hückel molecular orbitals of benzene, viewed from above the plane of the molecule. There will be two components to each lobe shown, one above the plane of the molecule and the other below the plane with an opposite sign

The total π electron energy is equal to $2(\alpha + 2\beta) + 4(\alpha + \beta) = 6\alpha + 8\beta$. This is to be compared with an energy of $3(2\alpha + 2\beta)$ for three localized π bonds. The delocalization energy is therefore equal to -2β, or about 150 kJ mol^{-1}.

Summary of Key Points

1. The variation principle states that the expectation value for the energy of any trial wavefunction will always be greater than the true ground state energy. By comparing the energies of a series of trial wavefunctions, it is therefore possible to arrive at the best approximation to the true wavefunction.

2. In calculating the electron wavefunction for a molecule, the nuclei can be treated as stationary because they move much more slowly than the electrons. This is known as the Born–Oppenheimer approximation.

3. Approximate wavefunctions for the bonding and antibonding states of H_2^+ can be obtained by combining hydrogen atomic wavefunctions centred on the two nuclei. This is known as the linear combination of atomic orbitals (LCAO) method.

4. Trial wavefunctions for H_2 can be produced either by putting both electrons into the LCAO orbital obtained for H_2^+ (the molecular orbital method), or by bringing separate H atoms together, and letting the electrons exchange places (the valence bond method). Neither method gives a satisfactory bond dissociation energy, but a wavefunction which is intermediate between the MO and VB wavefunctions provides much better agreement with experiment.

5. The LCAO method can be used to obtain molecular orbitals for any diatomic molecule from the appropriate atomic orbitals. Minimizing the energy leads to an equation known as the secular determinant, from which the best energies for the molecular orbitals can be calculated.

6. Hybrid orbitals, obtained by mixing of the 2s, $2p_x$, $2p_y$ and $2p_z$ orbitals on the central atom, can be used to predict the shapes of molecules, for example BeH_2, BH_3, NH_3, CH_4.

7. A satisfactory picture of bonding in the H_2O molecule can be obtained from either a MO or VB perspective.

8. The Hückel method has been used to obtain π molecular orbitals, and their associated energies, for butadiene and benzene. The π electron energy is found to be lower than it would be for localized π bonding, and this difference is known as the delocalization energy.

Problems

8.1. The overlap integral, S_{AB}, for the H_2^+ molecule-ion can be evaluated from the equation:

$$S_{AB} = \left(1 + \frac{R}{a_0} + \frac{R^2}{3a_0^2}\right)e^{-R/a_0}$$

Calculate a value for the overlap integral at the equilibrium separation, $R = 106$ pm. Use a value of 53 pm for the Bohr radius, a_0. What is the maximum possible value of S_{AB}, and when does it occur?

8.2. The molecular orbital for a heteronuclear diatomic molecule is $\phi = N(\psi_A + \lambda\psi_B)$, where ψ_A and ψ_B are normalized atomic orbitals. Find an expression for the normalization constant, N, in terms of the parameter λ and the overlap integral S_{AB}.

8.3. Discuss the relative stabilities of the following species in terms of bond order: (a) Li_2 and Li_2^+; (b) C_2 and C_2^-; and (c) F_2 and F_2^+.

8.4. The bond lengths of O_2^+, O_2, O_2^- and O_2^{2-} are 112, 121, 135 and 149 pm, respectively. Explain the variation in bond length.

8.5. Discuss the hybrid bonds present in the methyl radical, $CH_3^{\bullet}$, and explain where the unpaired electron is located. The methyl radical has a planar structure.

8.6. Show that the sp^3 hybrid orbitals of carbon, given in equations (8.57)–(8.60), are normalized and have zero net overlap with one another; that is, they are orthogonal. It can be assumed that the contributing atomic orbitals are themselves normalized and orthogonal.

8.7. The bond angle in H_2S is 92.2°. Estimate the proportion of sulfur 3s orbital character in the bonding orbitals.

8.8. Find the Hückel molecular orbitals and energies of the allyl radical, $\bullet CH_2$–CH=CH_2. What is the delocalization energy?

8.9. Use the Hückel approximation to calculate the π-electron energies of cyclobutadiene. Hence determine the electronic ground state of cyclobutadiene. What is the delocalization energy of this molecule?

8.10. The enthalpy change for hydrogenation of cyclohexene to cyclohexane is equal to -121 kJ mol^{-1}, whereas the enthalpy change for hydrogenation of benzene to cyclohexane is -209 kJ mol^{-1}. Calculate the stabilization energy of the delocalized π-electron system in benzene. Note: cyclohexene has a single C=C double bond.

References

1. P. W. Atkins, *Physical Chemistry*, 6th edn., Oxford University Press, Oxford, 1998, p. 397.
2. W. Kolos and L. Wolniewicz, *J. Chem. Phys.*, 1968, **48**, 3672; *J. Chem. Phys.*, 1968, **49**, 404.

Further Reading

J. Barrett, *Structure and Bonding*, Tutorial Chemistry Texts, Royal Society of Chemistry, Cambridge, 2001, chaps. 4 & 5.

D. A. McQuarrie and J. D. Simon, *Physical Chemistry: A Molecular Approach*, University Science Books, Sausalito, California, 1997, chap. 10.

P. W. Atkins and R. S. Friedman, *Molecular Quantum Mechanics,* 3rd edn., Oxford University Press, Oxford, 1997.

Answers to Problems

Chapter 1

1.1. Threshold wavelength = 451 nm.

1.2. Maximum velocity = 8.25×10^5 m s^{-1}.

1.3. (a) Kinetic energies of photoelectrons are 7.2, 4.4 and 1.5 eV. (b) Photon energy of neon radiation = 16.7 eV. This energy is sufficient to remove electrons only from the highest energy level of CO. These electrons will have a kinetic energy of 2.7 eV.

1.4. Wavelength of neutron beam = 2.47×10^{-10} m.

1.5. Wavelengths: (a) $\lambda = 1.23 \times 10^{-10}$ m; (b) $\lambda = 6.72_5 \times 10^{-11}$ m. Angle at which first-order diffraction peaks observed: (a) 34.6°; (b) 18.1°.

1.6. Wavelength of argon beam = 1.79×10^{-11} m. Diffraction angle is very small (4.75°) and diffraction is therefore unlikely to be observed.

1.7. (i) Momentum = 3.82×10^{-24} kg m s^{-1}; (ii) wavelength = 1.73×10^{-10} m; (iii) $k = 3.63 \times 10^{11}$ m^{-1}.

1.8. $P = 2\int\limits_{0}^{\infty} e^{-2x}dx = 1$. Hence, wavefunction is normalized.

Probability of finding particle between $x = -1$ and $x = +1$ is equal to $2\int\limits_{0}^{1} e^{-2x}dx = \left(1 - e^{-2}\right) = 0.865$.

1.9. Only function (ii) is satisfactory; the rest are not. Function (i) $\rightarrow \infty$ as $\pm x \rightarrow \infty$. Function (iii) equals ∞ when $x = 3$. Function (iv) $\rightarrow \infty$ when $x \rightarrow -\infty$.

Chapter 2

2.1. $\Delta E = 0.064$ eV. This value is slightly higher than the experimental value because the real potential well has finite walls, and this allows the wavefunction to extend slightly into the surrounding gallium aluminium arsenide.

2.2. For transition $n = 1$ to $n = 2$, $\Delta E = 3.54 \times 10^{-38}$ J; n(300 K) $= 4.19 \times 10^8$. Thermal energy $>> \Delta E(1 \rightarrow 2)$. Therefore quantum effects will not be observed.

2.3. Wavelength of light $= 352$ nm (for box length of 864 nm).

2.4. (a) $\Delta E = 1.38 \times 10^{-19}$ J; (b) wavelength $= 1.44 \times 10^{-6}$ m; (c) average number of π electrons between C_{11} and $C_{12} = 1.09$; (d) C–C bond order $= 1.54_5$.

2.5. Probability, $P = \dfrac{1}{4} - \dfrac{1}{2n\pi}\sin\left(\dfrac{n\pi}{2}\right)$. For $n = 1, 2, 3$ and 4,

$P = 0.091, 0.25, 0.303$ and 0.25, respectively. As n becomes bigger, P oscillates around the classical value of 0.25, with diminishing amplitude.

Chapter 3

3.1. Uncertainty in velocity, $\Delta v \geq 5.78 \times 10^6$ m s^{-1}.

3.2. $\dfrac{\Delta v_x}{v_x} = \dfrac{\Delta p_x}{p_x} \geq \dfrac{1}{4\pi}$.

3.3. $\Delta x \approx a_0$; $\Delta p_x \approx \dfrac{\hbar}{a_0}$; $\Delta x \Delta p_x \approx \hbar$.

3.4. (a) $\Delta x \geq 4 \times 10^{-4}$ m; (b) $\Delta x \geq 0.4$ m.

Chapter 4

4.1. $E = (\hbar^2 k^2)/2m$.

4.2. Zero point energies for HCl and DCl are 17.9 and 12.8 kJ mol^{-1}, respectively.

4.3. $k = 1549$ N m^{-1}. This value is slightly lower than the accurate value because the vibration of the molecule is not strictly harmonic.

4.4. Tunnelling current increases by a factor of 465.

Chapter 5

5.1. Normalization constant $= 1/\sqrt{\pi}$. Allowed values of α are 1, 2, 3, *etc.*

5.2. $\Delta E(J = 0 \rightarrow J = 1) = 8.41 \times 10^{-22}$ J; $\lambda = 2.36 \times 10^{-4}$ m.

5.3.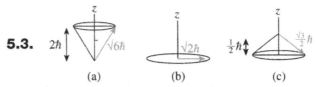

(a)　　　　(b)　　　　(c)

5.4. Minimum angle $= 45°$.

Chapter 6

6.1. $E_2 - E_1 = 10.20$ eV.

6.2. $r_{\min} = 4a_0$.

6.3. (a) $r = 6a_0 = 317$ pm; (b) $r = 1.76a_0 = 93$ pm; (c) $r = 10.24a_0 = 542$ pm.

6.4. $r = 5.24a_0 = 277$ pm.

6.5. $\langle r \rangle = 1.5a_0 = 79.3$ pm.

6.6. $E = -\dfrac{\mu e^4}{8\left(4\pi\varepsilon_0\right)^2 \hbar^2}$.

6.7. $r = 2a_0$.

6.8. The electron must be in a $2p_z$ orbital.

6.9. $\psi\left(\mathrm{d}_{xz}\right) = \psi_+ + \psi_- = N\left(\dfrac{r}{a_0}\right)^2 e^{-r/3a_0} \sin\theta \cos\theta \cos\phi = N\left(\dfrac{xz}{a_0^2}\right)e^{-r/3a_0}$;

$\psi\left(\mathrm{d}_{yz}\right) = \psi_+ - \psi_- = N\left(\dfrac{r}{a_0}\right)^2 e^{-r/3a_0} \sin\theta \cos\theta \sin\phi = N\left(\dfrac{yz}{a_0^2}\right)e^{-r/3a_0}$

Chapter 7

7.1. $\hat{H} = \hat{H}_A + \hat{H}_B$ with $\hat{H}_A = -\dfrac{\hbar^2}{2m_e}\nabla_1^2 - \dfrac{e^2}{4\pi\varepsilon_0 r_{1A}}$ and $\hat{H}_B = -\dfrac{\hbar^2}{2m_e}$

$\times \nabla_2^2 - \dfrac{e^2}{4\pi\varepsilon_0 r_{2B}}$. $\quad \Psi(1,2) = \psi_{1sA}(1)\psi_{1sB}(2)$.

7.2. $\Psi(1,2) = \dfrac{1}{2}\left[\psi_{1s}(1)\psi_{2s}(2) + \psi_{1s}(2)\psi_{2s}(1)\right] \times \left[\alpha(1)\beta(2) - \alpha(2)\beta(1)\right]$

7.3. $\Psi = \psi_a(1)\psi_b(2) - \psi_a(2)\psi_b(1)$. Interchanging (1) and (2) causes Ψ to change sign, showing that it is an antisymmetric wavefunction.

7.4. The effective nuclear charges for Na, K and Rb are $-1.84e$, $-2.26e$ and $-2.77e$, respectively.

7.5. Effective nuclear charge $= -2.36e$. The screening constant $= (3 - 2.36) = 0.64$, showing that the 1s electrons are not very good at screening the nuclear charge from one another.

7.6. The nuclear charge increases by one in going from Na to Mg, and the increased electron–nuclear attraction more than offsets the effects of inter-electron repulsion.

7.7. The transfer of an electron from a 4s to a 3d orbital will cause a net increase in inter-electron repulsion, which is sufficient to raise the energy of the alternative configurations above the ground state configuration.

7.8. Number of unpaired electrons in F, P, Ti and Co is 1, 3, 2 and 3, respectively.

7.9. Phosphorus has three unpaired electrons in separate $2p_x$, $2p_y$ and $2p_z$ orbitals. The extra electron in sulfur must pair with one of these electrons, and this causes increased inter-electron repulsion.

Chapter 8

8.1. At equilibrium separation $S_{AB} = (1 + 2 + \frac{4}{3})e^{-2} = 0.586$, Maximum possible value for S_{AB} is 1. This occurs when $R = 0$.

8.2. $N = \dfrac{1}{\sqrt{\left(1 + 2\lambda S_{AB} + \lambda^2\right)}}$.

8.3. (a) The bond orders of Li_2 and Li_2^+ are 1.0 and 0.5, respectively, and the ion is less stable than the neutral molecule. (b) Extra electron in C_2^- goes into bonding orbital, giving bond order of 2.5; C_2^- is therefore more stable than C_2. (c) In F_2^+, an electron is removed from an antibonding orbital, giving a bond order of 1.5. F_2^+ is therefore more stable than F_2.

8.4. O_2 has one electron in each of the antibonding $\pi_g^* \, 2p_x$ and $\pi_g^* \, 2p_y$ orbitals, and a bond order of 2. O_2^+ has one less electron, and therefore an increased bond order of 2.5. On the other hand, O_2^- has an extra electron which goes into the antibonding orbital, giving it a bond order of 1.5. Similarly O_2^{2-} has a bond order of 1.0. Thus, the bond length will be shortest in O_2^+ and greatest in O_2^{2-}.

8.5. The methyl radical has three sp^2 hybrid bonds linking the carbon atom to the three hydrogen atoms. The fourth electron is located in the remaining, non-bonding carbon 2p orbital, which is oriented at right angles to the plane of the radical.

8.6. For ϕ_1 we have:

$$\int \phi_1^2 \mathrm{d}\tau = \tfrac{1}{4}\int \left(\psi_{2s}^2 + \psi_{2p_x}^2 + \psi_{2p_y}^2 + \psi_{2p_z}^2 \right)\mathrm{d}\tau = \tfrac{1}{4}\left(1+1+1+1\right)=1$$

All cross terms, such as $\int \psi_{2s}\psi_{2p_x}\mathrm{d}\tau$, are equal to zero, and do not appear in the equation. A similar calculation applies to the other hybrid orbitals. Similarly, $\int \phi_1\phi_2\mathrm{d}\tau = \tfrac{1}{4}\int(\psi_{2s}^2 - \psi_{2p_x}^2 - \psi_{2p_y}^2 - \psi_{2p_z}^2)\mathrm{d}\tau = \tfrac{1}{4}(1-1-1+1) = 0$ and all the hybrid orbitals are found to be orthogonal to one another.

8.7. Contribution of the 3s orbital to the bonding orbitals is 3.7%.

8.8. The energies are: $E_1 = \alpha + \sqrt{2}\beta$, $E_2 = \alpha$ and $E_3 = \alpha - \sqrt{2}\beta$. The corresponding Hückel molecular orbitals are:

$\phi_1 = 0.50\psi_1 + 0.71\psi_2 + 0.50\psi_3$

$\phi_2 = 0.71\psi_1 - 0.71\psi_3$

$\phi_3 = 0.50\psi_1 - 0.71\psi_2 + 0.50\psi_3$.

Delocalization energy = -0.83β.

8.9. Energies are: $E_1 = \alpha + 2\beta$; E_2, $E_3 = \alpha$; $E_4 = \alpha - 2\beta$. Ground state contains two electrons in E_1, and one each in E_2 and E_3, with parallel spins. There is zero delocalization energy.

8.10. Stabilization energy = $(3 \times 121) - 209 = 154$ kJ mol^{-1}.

Subject Index

Lightning Source UK Ltd.
Milton Keynes UK
UKOW05f2325210216

268812UK00001B/22/P